Creative Wellsprings For Science Teaching

Second Edition

ALFRED DE VITO

Professor Emeritus of Science Education

Purdue University

CREATIVE VENTURES, INC.
P. O. Box 2286
West Lafayette, Indiana
47906

LIBRARY OF CONGRESS CATALOG CARD NO. 87-073548

De Vito, Alfred

   Creative Wellsprings for Science Teaching

      Second Edition

Copyright © 1989 by CREATIVE VENTURES, INC.
P. O. Box 2286, West Lafayette, Indiana 47906

All rights reserved. No part of this book may be reproduced in any form or by any means without permission in writing from the publisher.

Current Printing (last digit)

10 9 8 7 6 5 4 3 2 1

ISBN: 0-942034-06-6

Cover designed by Kathy Shuster
Art by Donna Acosta and A. De Vito

Printed in the United States of America

Preface

> "Only the brave. Only the brave should teach. Only those who love the young should teach. Teaching is a vocation. It is as sacred as the priesthood; as innate a desire as inescapable as the genius which compels a great artist. If he has not the concern for humanity, the love of living creatures, the vision of the priest and the artist, he must not teach."
>
> Pearl Buck

This quotation has served me well. Once I deemed it so vital I distributed copies to the students in my class. The anticipated responses were not what I had expected. The typist had inadvertently typed "Teaching is a vacation." My first day's introduction to the teaching profession alerted me to the fact that "Teaching is not a vacation."

MY FIRST DAY

(or Operation Slingshot)

Three thousand tear tissues were strategically distributed about the room. The wax begonia plants were appropriately wiped and glazed until they glistened. The gerbils, nearly deranged from the summer solitude, were given mouth-to-mouth resusitation to sustain them for another year's scientific investigation. It was "D" day and this was my first day at Lucy Grusom Elementary School. I was the new 2nd grade teacher. And I was ready.

Professor Thornside's voice still rang in my ears, "Start tough, show them no mercy, you can always let up as you and the year drones on." I had mapped out my strategy for the forthcoming, 186 hand-to-hand combat days with those pencil-swinging, bubblegum-chewing commandos. This would be the test. Would my infantry training stand me in good stead? Was it to be humanism or cannibalism? They could have it either way. In any event I was going to be ready. My strategy was fool proof. I would position myself by the window, my back to the enemy. The position had to be perfect. The sun's rays must strike me just right in order to cast the darkest shadow over the greatest area. When the children arrived, I planned to grab at my waist, huggingly trapping a portion of my elastic waist band. Having thus enmeshed my fingers in this elastic slingshot, I would slowly stretch this monster from my deodorized body. I would increase the tension until I had reached 7 on the Lucy Grusom Elementary School Richter scale. Then, I would release this inhumane juggernut, slingshot style, against my leathery body letting the sound reverberate against the cinderblock walls much like a mother beaver warning her young. At the same instance, I would whirl around, and in my best authoritarian tone, inquire, "Are we ready?"

A flashing, yellow light dancing across the window pane signalled the arrival of the school buses and their precious cargoes. I rushed to the window and positioned myself. The countdown had started. There was no turning back. My senses sharpened. Slowly the door opened. The little elves were invading the room. I was tempted to take at least one peek. Surely, the first ones in had to be my future, gold-star winners. Restraining myself I managed to maintain a suitable decorum. I could hear the soft squeegeeing of the deep tread of new sneakers gliding across freshly waxed floors. An occasional squeal of rubber punctuated the air as abrupt 90 to 180 degree turns were made as various students selected sunshine areas over eclipsed areas. Voices dwindled. It was working! At last I had living proof that those tired cliches meted out in methods courses, such as, "plan your work and work your plan" worked. Or so I thought, until some junior-sized fullback mistook me for a clothes tree and proceeded to drape a steamy, Kansas-City-Chief sweat shirt over me. A pedogogical grunt coupled with a fast two step resurrected from my famed physical education course set me free. I recovered my equilibrium despite the rapidly engulfing aroma of Johnson Baby Oil mixed with a tinge of borrowed musk oil. I knew the time was ripe. If only Mr. Rizzo, the principal, could be here to dim the lights and cast the beam from the overhead projector on me. It was at this moment that John Barrymore and I grew intensely close. A dizzyness swept over me. Suppose this self inflicted punishment stunned me? Suppose I passed out? Suppose I bled to death? Or suppose this lethal garment turned on me and strangled this total educational machine? Was it worth it?

A shaking hand snapped at the waist. These fingers clutched

anxiously at the material. The tug of war was underway. Exceeding the predetermined Lucy Grusom scale by a value of one, the release was initiated. The roaring slap could only be compared to an avalanche echoing down the length of Grand Canyon. With no visible signs of blood and only a dull ringing in my ears I spun around. No words left my mouth. My lips were galvanized shut. There sat 31 Lilliputian cherubs, several holding nosegays spiked with an occasional dandelion. One or two had gifts of fruit - damaged but still recognizable. Through slowly misting eyes I spotted a fig newton shyly held up as a peace offering. I melted. Gasping, I blurted out, "Welcome to Grusom!"

I struggled through the day and I only used 2,981 tear tissues. I wasn't sure if I had won or lost this battle until the closing bell. Then I overheard one chocolate-mustached cherub mumble, "He ain't going to be the best teacher in the world, but he'll do until one comes along."

. . .

I am still 'coming along.' This book is about 'coming along.' It grew out of reaction to a Thoreau comment: After he had his winter supply of wood cut, had a good fire going in his stove, was cosy and comfortable, and was reflecting his blessings, he found himself disturbed at the thought that someone might ask him, "What did you do while you were warm?"

What do teachers do after they become warm? I believe they reach out for new, innovative ways to improve and embellish their teaching. This reaching out can take numerous routes. None is more fruitful than the incorporation of creativity into one's approach to teaching. This book is about creative ideas which stimulate thinking.

This book stresses creative teaching as a means for improving the quality of science education for all children. A segment is devoted to educating the gifted in science; another segment is concerned with science instruction and its enhancement through the use of provocative question asking; the skill of model building is promoted; and, peripheral enhancements such as discrepant events, puzzlers, problems, and tenacious "think abouts" are included. Curiosity is fostered but a "rage to know" is promoted as a more desired accomplishment.

The summary desired outcome of this book is to provide teachers with opportunities to develop any single activity or experiment to where it can serve a greater variety of individuals, for greater challenges, for a longer period of investigation, and in a more provocative, challenging atmosphere.

Alfred De Vito

Nothing in the world can take the place of persistence. Talent will not; nothing is more common than unsuccessful men of talent. Genius will not . . . the world is full of educated derelicts. Persistence and determination alone are omnipotent. The slogan "press on" has solved and always will solve the problems of the human race.

Calvin Coolidge

Contents

SECTION I: WHAT IS SCIENCE?	page	1
WHAT IS UNIQUE TO SCIENCE?		3
THE METHOD OF SCIENCE		5
CONTENT SELECTION IN THE SCIENCE CURRICULUM		7
THE LANGUAGE OF SCIENCE		8
SCIENCE FORMULA		11
SECTION II: SCIENCE CORNUCOPIA - EXTENDING SCIENCE		13
APPROACH 1: THE MORPHOLOGICAL APPROACH		18
APPROACH 2: THE PROCESS APPROACH		27
ACTIVITIES TO ENHANCE THE PROCESS SKILLS OF OBSERVING, INFERRING, AND PREDICTING		
OBSERVING ACTIVITIES		44
INFERRING ACTIVITIES		54
PREDICTING ACTIVITIES		58
SUGGESTED INSTRUCTIONAL ACTIVITIES FOR		
CLASSIFYING		63
GRAPHING		72
SUGGESTED INSTRUCTIONAL ACTIVITIES FOR IDENTIFYING AND CONTROLLING VARIABLES AND TESTING HYPOTHESES		80
A TRIAL RUN FOR VARIABLE IDENTIFICATION		81

ACTIVITIES TO ENHANCE THE PROCESS SKILLS OF
HYPOTHESIZING AND EXPERIMENTING ... 85
OPPORTUNITIES FOR INVESTIGATION ... 92
APPROACH 3: THE IDEATION – GENERATION APPROACH ... 105
WHEN "SCIENCING," HELP CHILDREN TO... ... 114
WHAT MIGHT A PROFILE OF A SCIENTIFICALLY LITERATE INDIVIDUAL LOOK LIKE? ... 115

SECTION III: STIMULATING THINKING THROUGH SCIENCE INSTRUCTION ... 117

CREATIVITY ... 117
WHAT IS CREATIVITY? ... 118
WHO IS CREATIVE? ... 118
THE COMPONENTS OF CREATIVITY ... 119
ATTRIBUTES OF CREATIVE PERSONS ... 120
THE DIVERGENT-CONVERGENT MODEL ... 121
HOW TO IMPROVE YOUR CREATIVITY ... 123
CREATIVITY CAN BE DEVELOPED BY... ... 124
BARRIERS TO CREATIVE TEACHING ... 125

SCIENCE FOR THE GIFTED ... 128
WHO ARE THE GIFTED? ... 128
SCIENCE TAILORED FOR THE GIFTED ... 129
CHALLENGES FOR CHILDREN ... 131
EXAMPLES OF A BASIC SCIENCE LESSON EXTENDED TO ACCOMODATE THE GIFTED ... 132
STIMULATING THINKING THROUGH SCIENCE INSTRUCTION ... 134
CREATIVITY AND QUESTIONING ... 139
WHAT WOULD HAPPEN IF... ... 143

ASKING PROVOCATIVE QUESTIONS	144
ADDITIONAL SUGGESTED CREATIVE INVOLVEMENTS	145
HOW AND WHY QUESTIONS	146
MODEL BUILDING - THE FORGOTTEN PROCESS SKILL	147
MYSTERY CONTAINERS	150
AN INTRODUCTORY MYSTERY - CONTAINER EXERCISE	150
VARIATIONS OF MYSTERY CONTAINERS	152
UMBRELLIC SCIENCE MODEL	157
A MODEL FOR DISSOLVING	161
A MODEL FOR SPATIAL DEVELOPMENT	178
SECTION IV: DISCREPANT EVENTS, PUZZLERS AND PROBLEMS, AND TENACIOUS THINK ABOUTS	228
DISCREPANT EVENTS AND SCIENCE INSTRUCTION	228
PUZZLERS AND PROBLEMS	257
TENACIOUS THINK ABOUTS - DILEMMA SCIENCE ACTIVITIES (PLUS INDEX)	264
ENHANCING VERBAL INTERACTION BY LEARNERS THROUGH WRITTEN DILEMMA ANALYSIS	266
ATTRIBUTES OF DILEMMA ANALYSIS	268
THE PREPARATION OF DILEMMA SITUATIONS	270
AN EXAMPLE	271
A SAMPLE DILEMMA SITUATION	273
THE FINALE	344
INDEX	347

A BOOK IS A LOOK INTO THE PAST WHICH WHEN COMMINGLED BY THE READER WITH THE PRESENT CHARTS A PATH INTO THE FUTURE.

De Vito

Section I: What Is Science?

Science is a many splendored thing.

Science is a look at tomorrow, today.

Science is like a ride on a merry-go-round,
   it is a pleasant experience, but it
   doesn't get you anywhere!

Science is an excursion into the unknown,
   piloted by hypothesized inquiry arriving
   at theorization built on experimentation.

Science is a panacea for Pandora's box.

Science is a geometric kaleidoscope cascading
   us into infinity.

Science is commanding, demanding...it is a
   Lorelei whose call can be our elixir or
   our annihilation.

Science is all of the above and none of the above. These poetic descriptions are only facets of the "thing" we call science. Truly, science means many things to many people. The very word "Scientific" seems to carry with it some connotation as to that which is desirable, a hallmark of progress, something awesome, or an intellectual approach. Its usage is wide and varied.

Science to the scientist has a somewhat different meaning. The scientist thinks of science as a study. The scientist basically recognizes two types of science: pure science and applied science.

Pure science is sometimes referred to as the "cutting" edge of science.

It is that area of science where new, creative knowledge is added to the field. Applied science is often referred to as technology or the application of this "new" knowledge. While these two types of science differ greatly in basic value, they do complement one another. Sometimes these two aspects of science are referred to as the frontier and the interior of science. Pure science being at the frontier and technology being at the interior of science. By virtue of position, technology is not less valued, it is a vital component of that which we call science. Pure science is an intellectual activity. Technology is a practical one. Science deals with ideas, technology deals with things.

Another definition often offered for science is "Science is a body of classified, organized and systematized knowledge." Perhaps the greatest injustice that can be done to science is to regard it merely as a collection of facts or a body of organized knowledge and the practice of science as little more than the continued accumulation of such facts. Science deals with hard, inflexible facts, but it also has to do with very general ideas and abstract principles.

What then is science? It is a search for order in nature. Science is both experimental and accumulative. Its structure is composed of removeable blocks wherein stronger, newly discovered blocks are constantly replacing weaker blocks, further solidifying the structure. Science is an elusive mistress. When you think you have it mastered, it turns, in octopus fashion opening doors, closing doors, turning corners, and alluding to clues always challenging the courageous searcher to press on. Some have described the act of sciencing as a form of madness. The madness is the non-ending search for order in a seemingly endless maze of geometrically increasing avenues of search.

The kind of science one teaches usually is a reflection of one's accepted definition of science. While some fanciful definitions of science exist and some, such as this short excerpt from "Alice in Wonderland," carry a measure of the flavor of science, they may aptly be termed partial but insufficient definitions.

> "Are we nearly there?" Alice managed
> to pant out at last. "Nearly there!"
> the Queen repeated, "Why we passed it
> ten minutes ago!"

The study of science advances at such a rapid rate that by the time you have defined some aspect of it, it has moved on and is no longer what it once was. A more accurate and manageable portrayal of science may be described by a definition as follows:

> Science is a search for knowledge through experimentation; a search for knowing and understanding; a questioning of all aspects of the environment; and the collection and analysis of data and the interpretation of their significance. Science is a human undertaking, open ended and dynamic.

If this above definition of science is adopted by you, then the kind of science program you prescribe will be in concert with it. This text will endeavor to reinforce this definition of science.

WHAT IS UNIQUE TO SCIENCE?

Science education is a human enterprise. It is that experience wherein individuals strive to explain natural phenomena. This includes the on-going process of seeking explanations and understanding of the natural world, plus that which this process produces. Content is the product of science, but it of itself is not science. Science is both process and product.

Each science discipline has its particular approaches, tools, and methods for discovering and ordering information. However, common to all is the process of experimentation. Experimentation is the process of establishing evidence, the existence of something, or a solution of a problem. Experimentation is what is unique to science.

Necessary to the process of experimentation, and concomitant to the content acquisition, is the development of the processes of science. This involves training in order to acquire proficiency in the discrimination of observations, classification of observations and other information, the quantification of observations through graphic communication, the synthesizing and modification of explanations, and the establishment and testing of predictions against theories. Proficiency and success in experimentation is predicated on knowledge of these supporting processes of science.

As important as content and process acquisition are, of equal importance is the personal achievements that result from an individual's involvement in the act of sciencing. If science is to gain stature in the eyes of the practitioner, it must be promoted on the basis of what science can do for the individual through the personal use of scientific thought.

Individuals should acquire from a study of science such attributes as:

- the questioning of all things
- a consideration of premises, impinging variables, and the consequence of proposed action
- a spirit of demonstrative verification
- a search for data and their meaning, and
- a respect for logic, and the development of strategies for inquiry.

Science education should extend the individual's ability to learn, to relate, choose, vote, communicate, challenge, and respond to challenges so that the individual may live with purpose in the world of today and tomorrow and achieve pleasure and satisfaction in the process.

THE METHOD OF SCIENCE

SCIENCE METHODOLOGY?

> "I should see the garden far better," said Alice to herself, "If I could get to the top of that hill; and here is a path that leads straight to it—at least no, it doesn't do that—but I suppose it will at last. But how curiously it twists! It's more like a cork-screw than a path! Well this turn goes to the hill. I suppose—no, it doesn't! This goes straight back to the house. Well then, I'll try it the other way."

Science education should assist learners in acquiring the skills, the knowledge, and emotional adjustment needed by them to relate successfully to themselves and to the world around them. Learners should be afforded the maximum science education so that they can best manage their personal and collective lives to the greatest extent and also enable them to survive in the face of change.

Desirable outcomes of a science curriculum should concern itself with:

> The acquisition of scientific knowledge and the use of such knowledge in the solution of problems.
>
> The social arena wherein science, the scientist, and the general public interact on a scientific basis with the question of survival being paramount.
>
> And, the personal dimension wherein personal fulfillment, from the attributes of the application

of science, is reached.

> "Everyday I teach, I learn.
>
> **Each** day, I learn more than I teach."
>
> Author Unknown

How does one teach one to teach science? The obvious answer is "very slowly. With great care. Applying tender loving care. Instilling a burning desire to wonder. Creating a burning desire to want to find out what you want to know and then perhaps doing something worthwhile with this information once you have acquired it. Teaching strategies and the processes for investigating science through experimentation. Providing for the acquisition of a scientific attitude. Providing for content acquisition. Distinguishing from the domain of science that science most appropriate for a particular grade level and for various students whose capabilities require special attention. Providing instruction in the underlying structure of science. Relating the various psychologies of learning to science teaching. Mixing this in with excellent questioning skills. Alerting participants to the pluses and minuses of various textbooks and programs for instruction. Pointing out ways to integrate science into the total curriculum. Cautioning teachers of the various safety concerns relative to the teaching of science. Instructing teachers on techniques for testing in the sciences. Etc. Etc. Etc..."

To attempt to "squeeze" into any one book all that is germane to the teaching of science would be to perform a disservice. Any attempt to do so would reduce topics to mere vignettes. The task of teaching science is complex. The time spent in a science methodology course is short. And, the resources there are sometimes thin. In lieu of these constraining parameters it appears that some practical, despite all concerns for the ancillaries, priorities could evolve from the numerous statements of "How does one teach one science?" Undoubtedly your goals for your students would impinge on this prioritization. A suggested end goal might be "Students who exit your science class will be more creative, innovative, independent thinkers."

CONTENT SELECTION IN THE SCIENCE CURRICULUM

Generally speaking elementary school science should be exploratory in nature. The nature of the learner dictates the curriculum rather than the structure of the sciences. The curriculum in the primary grades should concentrate on the development of the basic process skills of science such as observing, classifying, measuring, communicating, inferring, predicting, and space/time relationships. These skills should be developed around "hands on" science activities that stimulate curiosity and enthusiasm, that capitalize on the creative nature of young children, and that foster continuous creativity. The acquisition of a science vocabulary, while important, should not be acquired at the expense of ideas. Ideas come first and the vocabulary of science later.

The vitality of science should be promoted by constantly presenting the challenges of science. Inquiry and the spirit of searching for answers should not be sacrificed at the expense of the tedium of science. This is not to mask the tedium of science. The tedium, and there is much of this in the quest for answers to problems in science, is tolerated best when the attributes of science are firmly anchored in the learner.

The capstone to any science instruction should be the capacity of an individual to relate sensitively, to think divergently, and to perform imaginatively in personal confrontations with people and ideas.

THE LANGUAGE OF SCIENCE

Science, as with almost every area of study, has its own unique language. Students of science speak, write, or work using terms such as ergs, watts, mass, density, specific gravity, foot pounds, etc. These terms are necessary to communicate in the sciences. How else could we state "Antogony recapitulates phylogeny" so succinctly? Vocabulary can be divided into two general areas. One deals with the mechanics of doing science. The other has to do with the structure of sciences and those terms about science.

Science is not immuned to the erosional process of precision in communication. The word generalization might be used when the term law is intended, or vice versa. Students use "Where are you coming from?" when they mean "Frame of Reference." Teachers use the term experiment when they mean an activity. The language of science has almost become meaningless. Greater precision is needed. Examine the following vocabulary words dealing with the structure of science. At least five descriptions are provided (all from reliable sources). Read all descriptions and check one selection under each heading which best matches your understanding of the term. How do your selections compare to your classmates' selections?

GENERALIZATION:

...a statement whose nature is dependent upon the facets of each unit of a body of data and which in some way describes a generalizable, or universal principle common to all.
...is the act or process of taking something as a whole and not limiting it to a precise application.
...is a broad, overall statement or conclusion.
...a conclusion grouping and applicable to the majority (or parts of the majority) of the elements involved.
...working out and stating a rule that is vague which applies for all.

HYPOTHESIS:

...is a tentative explanation which may or may not be valid.
...a tentative theory or supposition provisionally adopted to explain certain facts and to guide in the investigation of others.
...a proposition (or set of propositions) proposed as an explanation for the occurrence of some specified group of phenomena, either asserted merely as a provisional conjecture to guide investigation or accepted as highly probable in the light of established facts.
...an unproved theory tentatively accepted to explain certain facts or to provide a basis for further investigation.
...a formula derived by inference from scientific data that explains a principle operating in nature, implies insufficiency of presently attainable evidence and therefore a tentative explanation.

OBSERVATION:

...act of taking notice--gathering of data, as for scientific studies, by recognizing and noting facts or occurrences--the information or data so obtained.
...the act or an instance of viewing or noting a fact or occurrence for some scientific or other special purpose. The information or record secured by such an act. Something that is learned in the course of observing things.
...scientific scrutiny of a natural phenomenon, for experiment, verification or measurement and calculation.
...a perceived set of circumstances whose nature may be independent or interrelated.
...the act, habit, or power of seeing and noting, the act of watching for some special purpose study.

LAW:

...rule or principle that must be obeyed. A principle based on the predictable consequences of an act or condition.
...a general principle to which all applicable cases must conform.
...a rule of construction or procedure.
...observed regularity of nature.
...implies a statement of order and relation in nature that has been found to be invariable under the same conditions.

INFERENCE:

...is a derived conclusion from facts or premises.
...a truth or proposition drawn from another which is admitted or supposed to be true; a conclusion; a deduction.
...a statement of a probable relationship between two systems of either a parallel or continuous nature.
...the process of deriving the strict logical consequences of assumed premises. The process of arriving at some conclusion which though it is not logically derivable from the assumed premises, possesses some degree of probability relative to the premises a proposition reached by a process of inference.
...implies arriving at a conclusion by reasoning from evidence--the act of passing from one statement considered as true to another whose truth is believed to follow from that of the former.

MODEL:

...a system of symbols which represents a larger, or more complex system; the traditional purpose of which is to gain insight into the latter's nature at a preliminary state.
...an example for imitation. A miniature representation of a thing. Something held up before one for guidance.
...a pattern of something to be made or followed.
...a theoretical entity used to illustrate or explain a physical system.
...a preliminary representation of something, serving as the plan from which the final object is to be constructed.
...an abstract which offers a plausible explanation for phenomenon and which has programmed into it the element of prediction.

THEORY:

...a hypothesis which has undergone verification, and which is applicable to a large number of related phenomena.
...is a closely reasoned integrated group of fundamental principles intended to serve as an explanation for a group of phenomena.
...abstract body of concepts that serves as a basis of practice.
...a statement which explains, or lends order to some system of data; and which has undergone a relative amount of formal and informal testing.
...an explanation of certain phenomena, i.e., a series of hypotheses found to be consistent with one another and with observed phenomena but not yet proved by experiment.

ASSUMPTION:

...the inference that a fact exists, based on other known facts. Anything taken for granted.
...a statement made on the basis of probability, as an indirect proof which, is traditionally used to facilitate some conclusion formation of logical continuity, such as an inference or hypothesis.
...is a supposition, presumption or anything taken for granted. It is a supposed fact or statement or presumption of that fact.

...the supposition that something is true; a fact or statement taken for granted.
...the act of taking for granted or supposing without proof that a thing is true; a supposition; a postulate.

⚫ ⚫ ⚫

Arrange the described words in a hierarchy based on which terms you feel are subordinate to subsequent terms. For example, all of science begins with observation, therefore, observation would rank number one. What, in a hierarchial order, simple to complex, is the next step above observation? Continue on, filling the remaining steps.

```
*  8. _____                              8. model
   7. _____                              7. law
   6. _____                              6. generalization
   5. _____                              5. theory
   4. _____                              4. hypothesis
   3. _____                              3. assumption
   2. _____                              2. inference
   1. | OBSERVATION |                              1. observation  *
```

SCIENCE FORMULA

Learning is a do-it-yourself process. It can occur in any setting be it within a group or in a solitary surrounding. In either case, learning achievement is accomplished by an aggrandized solitary struggle for its internalization by the individual or individuals involved. This struggle can be a continuous, pleasureable process enhanced by the continuance of the act of learning. It does not have to be painful. The old adage of "Learning is a slow painful process" does not always apply. A substitute adage might well be "Everyday in every way I get better at this. The better I become at this, the more I like it. The more I like it, the better I become." It is difficult to teach science if you do not believe in it.

Teachers open doors. Hopefully not all the doors. Teachers turn up only one corner of a subject, leaving unturned corners for the students to discover and uncover.

The "I Like It" formula:

Work + work + more work = Success = strong self concept + work = even greater success.

Section II: Science Cornucopia - Extending Science

Elementary school teachers' comments relative to "Why they do not teach science or why they do not teach it as often or as well as they would like," can be summed up by three general statements. These are:

- little time
- little materials or equipment
- insufficient background in the content of science

These are legitimate concerns. However, they are concerns with solutions. The element of time is an elusive concern. In a sense we are all created equal with respect to time. While no one knows how long one will live, each day for each of us contains the same amount of time - twenty-four hours. What we do with time is the important thing. If one wants to do something, one generally can find the time. Time for science in the elementary school may well be a priority concession based on one's attitude towards the value of science. It is difficult to find time for science if one does not believe in its usefulness.

What is unique to science? What is so valuable about science instruction that if one neglected to teach science to elementary school children, one would be remiss in one's duty? An examination of "science" would reveal numerous assets which can accrue from a study of science. Paramount to all these assets is the process of experimentation. Experimentation, while not relegated solely to science, is most elegantly contained within its domain. Experimentation is unique to science. No other area of the curriculum provides such access to this process. No other area does it as well or with such profundity.

A response to this grandiose uniqueness might be, "Alright, but what is so great about experimentation? Who needs it?" Experimentation is a way to solve problems. It is a strategy for searching out solutions to questions. Concomitant with the arrival at a solution to a problem experimentation engenders a valuable by-product. This by-product is the recognition, by the practitioner, of the power within an individual to recognize a problem, state the problem, and to design some procedure to solve a problem or to answer a question. Each engagement, buttressed by the practitioner's successes, enhances a powerful, positive, self image towards one's ability to resolve future problems or to answer future questions. If there is a "joy" to sciencing, it is this engagement and resolution of the problem or question and the reinforcement of the notion of "I did it." Self-concept improvement for children is a highly desirable trait transcending science-knowledge acquisition. Recognizing this asset of self-concept growth, teachers find time for science instruction.

Lack of materials, while a strong concern, should not be a major concern. In the elementary school some basic, non-teacher fabricated materials are needed, for example, timers, pyrex glassware, heat sources, measuring devices, and magnification devices. These must be purchased. Excluding these, the remaining materials and equipment can easily be acquired. Balances and some measuring devices can be fabricated. Substitutions for the numerous items necessary for science instruction are available in local hardware stores, grocery stores, department stores, toy counters, attics, basements, salvage yards, and numerous other places.

The single most frequently quoted retardant to the teaching of science in the elementary school is the feeling of personal inadequacies relative to knowledge of the "content" of science. This is certainly a legitimate concern. There is no substitute for someone who knows everything about everything. If one possessed this talent, there is still no guarantee that one could translate this knowledge into teachable acts for children. And, of course, the converse of this is true. One cannot teach what one does not know. One can never play down the acquisition of knowledge in the pursuit of improved teaching. Attendant to this, one can never learn all there is to learn. Somewhere between these stages of "all knowing" and "not knowing," one must teach as best as one knows how while pursuing additional knowledge. This sometimes requires a cavalier attitude wherein one states to oneself that one does not need to know everything, probably will not ever know everything, but despite this, one can lead children _to_ and _through_ viable learning sessions.

The teaching of science in elementary schools is more often than not relegated to student involvement with activities. These activities usually support or reinforce an area of science being studied. The entire class does the same activity, in the same manner, with almost the same anticipated results. The activities are most likely hands-on involvements, more "doing" than searching for questions and answers.

Activities can pave the way to experimentation. This extension does not happen as often as it might. Perhaps this is because teachers may not be aware of techniques to expand basic activities. Both horizontal (breadth) and vertical (depth) explorations can emanate from any basic activity. Three approaches are suggested here to assist teachers in generating a plethora of ideas to stimulate provocative, question-answering searches. Not all ideas so generated are great. Some will appear trivial and useless. This is part of the generative act. However, most ideas can move activities forward as experiments along with stimulating creative thought.

> "The chessboard is the world, the pieces are the phenomena of the universe, the rules of the game are what we call the laws of Nature. The player on the other side is hidden from us. We know that his play is always fair, just, and patient. But we also know, to our cost, that he never overlooks a mistake, or makes the smallest allowance for ignorance."
>
> Thomas Henry Huxley (1825-1895)
> "A Liberal Education"

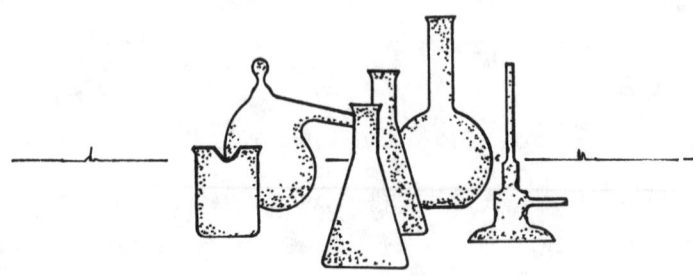

Most teachers are constantly reaching out for new, innovative ways to improve and embellish their teaching. This reaching out can take numerous routes. None is more fruitful than the incorporation of creativity into one's approach to teaching. Three approaches are presented here. These approaches are concerned with the generation of creative ideas which stimulate thinking. Proficiency in using one or more of these three approaches should enable the practitioner to become a "generator" as opposed to being a "duplicator" in science instruction.

By "generation" it is intended that teachers be able to take a basic activity and expand this into multiple activities and/or experiments. "Duplicating" refers to the replication of that which is described by another and, carrying it forth only as far as the author delineated it. And, letting it end there.

To possess the ability to develop any single activity to where it can serve more individuals, for greater challenges, for a longer period of investigation, and in a more provocative experimental atmosphere is a highly desirable teaching trait. Three approaches can assist in maximizing one's generation of ideas. These approaches can be used in isolation or in any combination. However, the use of all three approaches compounds the results.

APPROACH 1: THE MORPHOLOGICAL APPROACH - Idea Generation by Analyzing the Components of the System

Science activities can be expanded by following these steps:

Analyzing the system

- ask yourself, "What is involved?"
- list these components

Asking yourself "What and how" the components of the system can be manipulated

Establishing some priority for implementing the manipulations of the various components

AN EXAMPLE:

A time-tested activity which has received wide acclaim in science instruction is the "Wandering Spool." A traditional description of this activity might appear thusly:

Topic: Machines

Activity: How to make a "Wandering Spool"

Materials: wooden spool
two pieces of soda straw (one ten cm long and one straw shorter in length than the diameter of the spool)
thumbtack
plastic bead (from five mm to one cm in diameter)
rubber band (the length of the rubberband should be shorter than the height of the spool)

Procedure: assemble as shown in the sketch

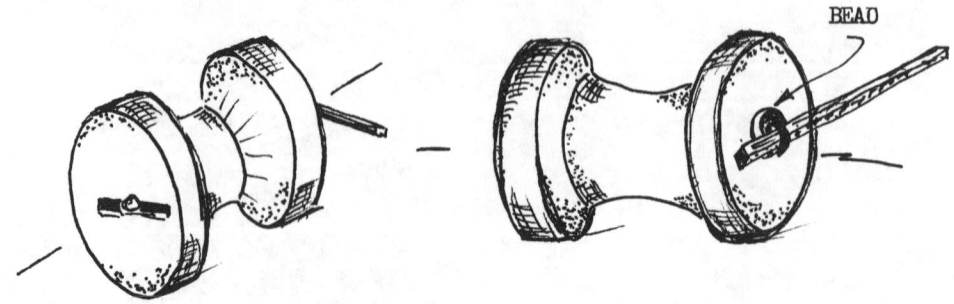

This is a great activity. Children love it. Interest is usually high. This activity moves. Almost anything that moves is an attention getter. Teachers like it because it is an economical and manageable activity. This means there is a lot of science mileage in this activity at very low expense. Numerous science concepts are involved, for example, energy, propulsion, work, friction, etc. Is this it? Is this where one stops? Perhaps. Perhaps not!

Three questions can serve as an indicator for the appropriateness of this lesson in terms of meeting the needs of all the members of your class.

    Ask yourself: <u>Should</u> everyone in the class be doing this activity?

                   <u>Could</u> everyone in the class do this activity?

                     <u>Would</u> everyone in the class want to do this activity?

If the answer is "yes" for all three questions, then there is a good chance that you are not meeting the individual needs of all the students in your class, particularly the gifted students. If the gifted children in your class <u>would</u> like to do something in science and <u>could</u> do that something, by virtue of some exceptional talents, it probably <u>would</u> not be suitable for the remainder of the class. Thus, to accomodate this diversity of needs, it becomes a highly, desirable teacher trait to be able to expand on basic activities providing a spectrum of involvements stemming from the same "root" or basic activity.

<u>USING THE MORPHOLOGICAL APPROACH FOR IDEA GENERATION</u>

    <u>Step A</u>: What is involved? In this particular activity the following is involved:

                spool
                soda straws
                plastic bead
                rubber band
                thumbtack
                surface area

Step B: How and what components of the system can be manipulated or varied? Think of as many ways as possible to manipulate, change, vary, etc. each component on your list permitting expansion on the basic activity, for example:

spool

- size of the spool
- mass of the spool
- shape of the spool
- container (for example, plastic containers vs wooden spools)

soda straw

- substitute wooden sticks (match stick devoid of the head), tongue depressors, swab sticks, doweling, etc.
- length of the straw
- number of straws

plastic bead

- different materials, for example, wooden beads, metallic beads, etc.
- size of the bead
- number of beads
- beads with varying diameter holes drilled through the bead

rubber band

- length of the rubber band
- thickness of the rubber band
- width of the rubber band
- number of rubber bands
- altering the rubber band to change the friction rate as it unwinds (for example, talc powder or some lubricant to reduce the friction of the rubber band)

thumbtack

- substitute a small nail, map pin

surface area

- altering the surface on which the spool travels, for example, waxed wood surface vs unwaxed surface, concrete, linoleum, various materials such as corduroy, silk, cotton (fastened down over wood or other material)
- change elevation (flat surface vs an inclined surface)

Step C: What is your priority for investigating any of the above?

This prioritizing may be predicated on your immediate goals. At some point in your instructional assignments based on grade level considerations, limitations of materials, etc., you may elect to perform a demonstration. This is a controlled performance to present some pre-selected phenomena. The results, barring any unforseen events, are generally known well in advance of the performance.

Or, you may wish to do what is aptly labeled "activities." Using the "Wandering Spool" idea, you may simply have each child construct one, operate it, and describe what was observed. Together you may search for an explanation of what was observed.

Greater prioritizing is necessary when you chose to involve children in experimentation. An experiment usually seeks an answer to a question that arises from some observable phenomena. It involves the following:

- the statement of a problem usually in the form of a hypothesis
- identification of variables that have a bearing on the stated hypothesis
- the formulation of strategies compatible with the stated hypothesis
- controlling the variables
- collecting and interpreting the data
- the conclusions

Use of the morphological approach contributes little to the act of demonstrating science or to the classroom science activity where only one outcome is the focus of the involvement. The morphological approach contributes most when you are looking for the most. This approach, wherein components are listed and manipulative suggestions stated, is fruitful of generating numerous, expanded activities. Reflecting on each stated variable, numerous questions can be raised. These can be restated in the form of hypotheses. This sets the stage for experimentation. Numerous questions! Numerous statements! Numerous experiments!

SOME SAMPLE EXPANSIONS:

Spool

. size of the spool

Question: If I increase the size of the spool will the spool travel further?

Statement or Hypothesis: If I increase the size of the spool, the large spool will travel further than a smaller spool.

Action: Set up two spools. Keep everything constant except the size of the spools. Wind them up. Make an inference. Release them. Collect the data by measuring the distances traveled. Interpret the data. Make conclusions.

This investigation is a good one. It is not an easy one. The most common error is to compare the movement of two, dissimilar-sized containers and not to equate the masses of the two containers. This results in the collection of meaningless data. The results cannot be clearly attributed to the differences in size or to the differences in the masses of the two containers. Everything should be kept constant and only one variable manipulated. If this constraint is observed, any differences can then be attributed to that variable which was manipulated.

It is difficult to get two spools of different size that possess the same mass, or to get two spools of the same size which have differing masses. Spools can be weighted. Once the additional mass, needed to equate two spools or to dis-equate two spools, is determined, coiled soft, lead/tin solder can be shaped around the spool and gently hammered flat. Make sure the coiled solder does not interfere with the rolling motion of the spool.

Continuing, spools of graduated sizes, all of similar masses, now can be compared commensurate with some previously stated hypothesis.

- mass of the spool

    Question: Will varying the mass effect the distance a spool travels?

    Statement or
    Hypothesis: If the mass is decreased, the spool with the least amount of mass will travel further.

    Action: Take two spools of the same size and vary the mass of one. <u>Keeping everything else constant</u>, crank them up. Make an inference. Release them. Measure the distances traveled. Interpret the data. Make conclusions.

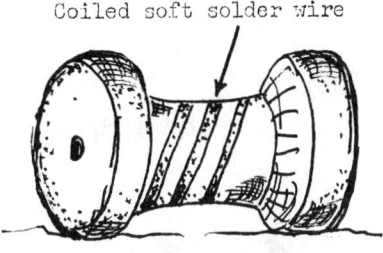

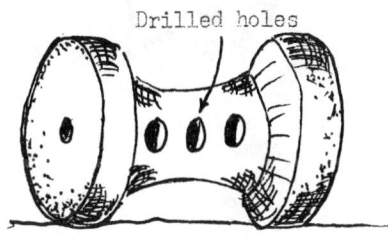

To further adjust for mass differences, small holes may be drilled into the shank of the spool and filled with lead shot, iron filings, etc. and then plugged with wax, clay or simply, glued closed.

- shape of the spool

    Question: Does it make a difference if I use a regular sewing spool or a spool cut from a broom handle (and drilled through)?

    Statement or
    Hypothesis: I believe that a sewing spool that has rimmed edges on which to roll will travel further than a spool cut from a dowel or broom handle which has a uniformly, smooth-rolling surface.

    Action: Assemble the two spools (sewing spool vs dowel or broom-handle spool). <u>Keep everything else constant</u>. Crank them up. Make an inference. Release them. Measure the distances traveled. Interpret the data. Make conclusions.

Wooden sewing spools are becoming increasingly more difficult to obtain. One inch or one and one quarter inches doweling cut in one and one half inches length segments, drilled through the center make excellent spools. Two smaller spools combined to equal one larger spool can be compared as to performance. The spool's cylindrical surface can be grooved to present alternate shapes. Any variations can be compared, one to another, to determine what effect these variables have on the final results.

Various cylindrical objects can be used as spools. Empty soft drink cans with holes punched in the top and bottom make excellent, inexpensive, readily-available spools. Empty plastic, vitamin pill bottles and pharmaceutical containers can be used to make small spools. Plastic, two-liter, soft drink containers can be used to make super spools. A nail, a nail punch, or an awl can be used to make holes. One should exercise great care punching holes in all materials. One should be particularly careful when punching holes in curved plastic surfaces. Keep your hands clear at all times when exerting pressure with sharp instruments. Another word of caution is necessary, do not use a heated object such as a nail to melt one's way through a plastic material. Noxious gases may be given off.

When the spool is manipulated, keeping the mass equal will always be a concern and a necessity. This is true when one considers the following manipulations:

- container

    Question: Can containers be substituted for spools and will they operate in much the same manner?

    Statement or Hypothesis: A container arranged in the same manner will operate in a fashion similar to a spool.

    Action: Any plastic bottle, 35 mm slide film container, pill container, etc. can be fashioned into a "wandering spool." Punch holes in both ends, assemble, keep everything the same except the container. Make an inference. Wind it up. Release it. Measure the distance it traveled. Record the data. Interpret the data. Make conclusions.

From a simple basic activity, considering manipulation of only the spool, we have generated four additional investigations. This same procedure can be replicated when one considers variations of the soda straw, the plastic bead, the thumb tack and the surface area. These considerations will generate at least fourteen more involvements. And, this is not the sum total of generations. The "wandering spool" seems to generate additional involvements in response to operational observations.

Some concerns worthy of investigation are:

- How to make it go in a straight line. Or a curved line.

- How to construct a multiple spooled device with single or independent spool action.

- How to handle the friction problem or lack of friction.

- How well does it travel up an inclined plane? What happens when the angle is increased?

A note of caution...

Nothing works. Nothing works well. Nothing works one-hundred percent. Nothing works like authors say they do. Not every child will succeed with the construction and manipulation of the "wandering spool." Manipulation generates problems. Frustration, aggravations, and consternations are all part of it. It is the arrival at solutions to problems that intrigues us.

When dealing with the "wandering spool," one of the most common problems is that when excessive energy (winding) is put into the system, and the spool is released, it spins and gyrates in place. The mass of the spool usually is too light for the amount of energy wound into the system. And, the spool dissipates this energy in the easiest manner possible. This problem is correctable.

CURIOSITY IS NOT ENOUGH. BETTER A RAGE TO KNOW.

APPROACH 2: <u>THE PROCESS APPROACH</u> - Generation by Utilization of the Processes of Science

This approach can augment the morphological approach. Or, it can be used in isolation. The process approach is rooted in the processes of science. The primary objective of the process approach is to assist children in acquiring competencies in the skills of science. Emphasis is on what a scientist does in the pursuit of understanding science. This is in opposition to emphasizing the content of science as accumulated by scientists.

The process approach like the morphological approach is not particularly useful when presenting science through demonstrations. It is more useful when involving children in science activities. However, it is <u>most</u> useful in the act of experimenting. In fact one cannot conduct an experiment without the processes of science. The process skills of science have been identified as follows:

>Observing
>Inferring
>Predicting
>Classifying
>Communicating
>>Graphing
>Using Numbers
>Space/Time Relationships
>Controlling Variables
>Hypothesizing
>Experimenting

The hierarchy of process skills can be thought of as a map whose destination is "experimentation." Experimentation is the capstone of science and each process skill a part of the warp and woof of science. Proficiency in each process skill is necessary to adequately carry out an experiment. Acquisition of any subset of these skills is insufficient for involving oneself in experimentation.

Primary grade children will be able to attain proficiency with the skills of observing, inferring, predicting, classifying, communicating, graphing, using numbers, and space/time relationships. Primary children experience difficulty with hypothesizing and controlling variables. Intermediate grade children, however, can and do understand all the process skills. And, of course, gifted children, at any grade level, violate all stated rules of learning and constantly surprise us with their breadth and depth of search and learn.

Teachers who are searching for additional means to expand science through the generation of additional ideas which might stem from a basic activity should strive for mastery of all the processes of science. A summary explanation for each process is provided.

OBSERVING

Science is rooted in observation.

Instruction in observation is not finite. It is continuous. It goes on forever.

Instruction in the process of observation starts with experiences in an awareness of the use of all the senses. Necessary precautions should be stressed relative to the senses of smell and taste.

Most observations are made of objects in a static state. This description, for all practical purposes, will not change over time. A piece of chalk described today will have the same characteristics, traits, or attributes as it will have tomorrow. However, placed in another situation where a chemical or physical change is introduced, another set of observations is generated. This set of observations describes a dynamic set. For example, a piece of chalk placed in dilute HCl (or vinegar) will react and an entirely new set of descriptions is engendered. This action will finally resolve itself and return to a static state. Thus, in essence, we have three sets of descriptions, before (static or equilibrium), during (dynamic or disequilibrium), and after (a return to static or equilibrium).

Another dimension of instruction in the process of observing is the recognition of comparative observations. This is where a base or cadre of attributes, characteristics, or properties of one object is compared to another. For example, I am 5' 8" tall, weigh 175 lbs., and 25 years old. You are 5' 3" tall, weigh 122 lbs., and are 19 years old. I am taller, heavier, and older. Quantitatively, these differences should be delineated. I am 5" taller, 53 lbs. heavier, and 6 years older. This leads to rank ordering based on any one attribute, for example, shortest to the tallest, etc.

Instruction in observation should include an awareness of direct observations as well as indirect observations. Direct observations are more commonly used than indirect observations. And yet, in science most observations and measures are obtained indirectly. A model for the interior of the earth as constructed from indirect measures obtained from seismic wave recordings is an example. The mystery box concept is a very helpful device in providing experiences for the gathering of indirect observations. A mystery box is simply an opaque container into which an object or objects is placed.

The box is sealed shut. From external manipulations, one is to deduce what the properties of the object inside the container are and, if possible, identify the object.

The assemblage of data from direct and indirect means moves naturally into the construction of models. Science is replete with models, for example, model of an atom, model of the earth, the "S" curve model, the normal-curve model, the rock-cycle model, etc.

OBSERVING ACTIVITIES...................... page 42

INFERRING

An inference is an interpretation of one or more of the five senses. For example, an unshelled peanut exhibits a two-nodule configuration. One cannot see that there are two peanuts enclosed, but from the shape, rattling noise, etc., one could infer that there are two peanuts enclosed within the peanut shell.

INFERRING ACTIVITIES...................... page 54

PREDICTING

Using the above example, if one had opened 100 or 1,000 double-noduled unshelled peanuts, and the accumulated data shows that 92% of the time there are two peanuts contained within, then the next time one picks up a two-noduled unshelled peanut one does not infer, one predicts. Predictions are based on accumulated data. Science is primarily concerned with change and relationships between changes. Scientists observe, record, and analyze these changes and relationships. From such analyses they then make predictions regarding other events or changes. Prediction is one of the most important functions of science.

        PREDICTING ACTIVITIES..................... page 58

CLASSIFYING

Classification is an extension of an observation. Classification initially is based on properties which are directly observable. All classification systems are arbitrary. Similarities and differences are noted. From the array of similarities and differences decisions are made to dichotomously separate these. Instructionally, it is suggested that all items being classified be tallied and identified as elements of the set. A flow diagram of dichotomous separations using in the left member separation the positive statement such as spherical; the right hand member of the subset in question should be expressed in a negative statement, for example, "not spherical." The flow diagram for any set of elements should be carried through until single subset categories exist for each element of the set. The accumulative description provides, for the system created, an "operational definition" or a definition with which one can operate with within the confines of the constructed classification system for each element of the set.

        CLASSIFYING ACTIVITIES..................... page 63

COMMUNICATING

Communication is the heart of science. Communications should be accurate, complete, concise, and devoid of impressions. Three general types of communication exist in science; oral, written, and pictorial. Consistent with good language art involvements, the scientist in oral and written expression strives for unity, coherence, and emphasis.

Pictorial scientific representations are principally confined to graphing.

COMMUNICATING ACTIVITIES................... page 72

GRAPHING

A graph is a device for communicating a relationship between two variables. Graphs consist of scales on two axes that are usually equal. The independent variable or the manipulated variable is plotted on the "X" axis and the dependent variable or responding variable is plotted on the "Y" axis. Both axes usually start at zero. A title should communicate the purpose of the graph. Interpolation is the process of finding information that is within a measured range. By interpolating, you make predictions of values that have not been included in the data on which a relationship has been established. If the observed values do not reveal a regular pattern, you would have no basis for an interpolation.

Predictions can be made from data wherein extrapolations are made. Extrapolation is the process of predicting relationships beyond the range of observations.

GRAPHING ACTIVITIES...................... page 72

USING NUMBERS

Skill in measuring requires not only the ability to use measuring instruments properly, but also the ability to carry out calculations with those instruments.

Precision in measurement refers to the agreement among observed values in repeated measurements. Accuracy, on the other hand, refers to the agreement of the observed measurement with the actual or true value. Actual or true value may be defined operationally as the measurement that is made as carefully as possible with the best measuring instrument that can be made.

Instruction in using numbers should include the ability to estimate.

Some quantities may be measured with a single instrument (ruler) and no computation required. Other measures may be made with a single instrument, but a mathematical operation is required to obtain the desired measure. Measurements of this type are called indirect measurements.

Measurements of speed, density, and heat are examples of derived quantities. To obtain measures of these quantities, two measuring instruments are used and mathematical operations are carried out to obtain the measure of the quantity. Derived quantities are expressed in derived units.

If something moves, quantitatively express this in how far, how fast, and in what direction it moved.

USING SPACE/TIME RELATIONSHIPS

If something moves it is reoriented in space, for example, a ball traveling down an inclined plane. It moved or rolled so many cm per second. This movement is observed and described as a space/time relationship. The space/time relationship is an important aspect of science. It permits us to describe motion in space related to time.

CONTROLLING VARIABLES

Things that change naturally or that can be changed on purpose are called variables. Scientists prefer to conduct investigations involving only two variables at a time. Usually one of the two variables is deliberately changed. This is called the independent or manipulated variable. Any changes in the other variable identified as the dependent or responding variable are studied. And when possible they are measured. All other possible variables are not allowed to change. Therefore, it is construed changes in the dependent variable must change from changes in the independent variable.

USING CONTROLLING VARIABLES ACTIVITIES...... page 80

HYPOTHESIZING

Characteristics of a hypothesis:

    tentative explanation

    can be used to predict events

    should be testable

    should be consistent with the observations

Some observed conditions:

    only one variable is manipulated at a time - other variables should be held constant

    the procedure to be used to vary the manipulated variable (independent) should be specified the variable expected to respond (dependent variable) to the experimental manipulation should be <u>specifically</u> named

the results expected in the event that the hypothesis is true should be specified

USING HYPOTHESIZING ACTIVITIES............... page 80

EXPERIMENTING

An experiment usually seeks an answer to a question through an investigation of some observable phenomena. Experimentation involves:

the delineation of a problem

the construction of hypotheses

the identification of pertinent variables

the organization of a design compatible with the formulated hypotheses

the controlling of variables

the act of collecting data

interpreting the data, and

summarizing the conclusions in light of the problem

USING EXPERIMENTING ACTIVITIES............... page 85

A discussion of idea generation stemming from the process approach could have continued using the "wandering spool" as the basic or "root" activity. This might have become tedious belaboring the same activity. Nevertheless, the implementation of this approach could have increased the number of spool expansions even further. The selection of a new activity gives credance to the flexibility of the application of these approaches to idea generation.

AN EXAMPLE:

Another time-tested activity which is still widely acclaimed is the "Glass Jar Garden." This activity can be described thusly:

    Topic: Seeds

    Activity: Planting Seeds

    Materials: Clean glass jar
                several lima bean seeds
                paper towels
                vermiculite
                water
                sunlight

    Procedure: Fold the paper towel into a cylinder. The height of the paper cylinder should be somewhat lower than the height of the jar at that point where the jar tapers inward. Place the paper cylinder in the jar. Expand the cylinder so that the paper hugs the glass wall. Inside the cylinder, half fill the jar with vermiculite. Pry the paper cylinder away from the glass, slip in the seeds, positioned as desired. Water the seeds by saturating the paper toweling. Place in sunlight.

This activity is a good, science activity. The seeds are planted. They grow. And, there is a nice feeling of accomplishment on the part of the participant. Sometimes it ends here. More can be garnered from this activity. The morphological approach suggests consideration of the various components such as the seeds, the volume and other properties of the container, the liquid being used, the amount of light, etc. The process approach provides a somewhat different facet to idea generation. Each science process skill should be evaluated in light of what it might contribute to the expansion of the basic activity. Let's consider a few.

<u>Observing</u>

Use all the five senses. Enumerate as many properties of the lima beans as are observable to you, for example:

> quantification of the mass
> quantification of the lima bean's dimensions
> hardness of the lima bean's exterior
> odor* of the lima bean
> taste* of the lima bean
>
>> *cautions - Not all things observed should be smelled or tasted, stress caution when necessary
>
> visual comparative descriptions of the bean in the dried state, soaked overnight in water, and growth over time
> visual description of a segmented lima bean
> textural description

Inferring

How many inferences can you make about the lima bean seed? For example:

> The lima bean with the greatest mass will grow to the greatest height.
> The fertility of the dried lima beans is uneffected by radical temperature changes while in storage.
> The size of a lima bean is based on the number of beans produced by a single plant.

Predicting

Predictions are based on data interpretation. This may limit the application of predicting unless the activity being dealt with is based on past experiences. If previous experiences with seeds exist, predictions may be made thusly:

> With any number of seeds planted, only eighty percent of that number will grow.
> The plotted growth of the lima bean will trace the normal "S" curve.
> The lima bean will grow the fastest in the first three days of growth.
> Once the growing lima bean is allowed to dry out, no amount of water will resurrect it.

Classifying

Classification based on similarities and differences almost requires a number of different varieties of seeds. Lima beans, however, can be classified based on size, mass, and overall dimensions.

> size
> mass
> color variations (if any)
> wrinkled vs unwrinkled

Communicating (graphing)

Data becomes more meaningful and convenient when tallied and graphed.

Some examples are:

> Using forty or fifty measures, measure the longest plane of the beans. In appropriate units, arrange the data from smallest to the largest measure and record how often each measure occurs.
> Graph the recorded growth of the plant after a period of time.
> Graph the amount of mass increase of the plant over a period of time.
> Graph the amount of growth of various parts of the plant over a period of time (for example, the stem, the leaves or a leaf, the root system).

Using Numbers

Numbers will be used in quantifying the following:

> temperature
> amount of water given to the plants
> mass
> measurement of growth
> units of time, for example, days, months, etc.
> graphing processes

Space/Time Relationships

If there is growth over a fixed period of time, we have a Space/Time consideration. Some examples are:

> the growth of the seed
> the growth of the plant
> growth of various segments of the plant
> percolation of water through the soil
> evaporation of the water from the soil

Controlling Variables

Variables are changes that occur naturally or which can be induced. Ideally, in an experiment all variables are held constant except one which is manipulated either naturally or induced by the investigator. Any changes or responses, compared to a control, may thus be construed as a result of the manipulation of that variable.

Experimentation, as conducted with young children, often results in ineffective, collected data due to the inability of the children to always recognize all the variables that impinge on the results. This may render the results useless. Once identified, variables, except for the manipulated variable, must be held constant. These concerns should not deter experimentation by children. Neglected or inaccurately interpreted variables become fewer with practice.

Some often overlooked considerations for seed growth that might effect the outcome are:

>sunlight
>air circulation
>humidity
>amount of vermiculite
>depth of seed planting in relation to the amount of vermiculite
>orientation of the seed within the container

Hypothesizing

Simply stated, a hypothesis is a testable statement relative to an observed phenomenon that invites a search for a solution to a problem. Many questions usually can be rephrased as hypotheses. Some sample hypotheses are:

>Seeds grow best when watered with rain water as opposed to tap water.
>Seeds with the greatest initial mass will produce the largest plants.
>Seeds planted at greater depths grow better than those planted at shallow depths.
>Seeds soaked overnight prior to planting do not grow as well as dry seeds.

Experimenting

Experimentation is the _analysis_ (reducing ideas to their component parts), the synthesis (the assemblage of diverse parts, ideas, etc. to reveal new patterns), and the application of comprehended knowledge to solve a problem. An experiment is a hypothesis in action. It is a search for an explanation to a problem. A problem well stated is half solved.

The science process of experimentation embodies most, if not all, of the processes of science. Sciencing starts with observation and moves up to experimentation which is the pinnacle of the science process hierarchy. While experimentation starts with observing, it involves inferring and/or predicting, hypothesizing, and the controlling of variables. The process of classification may be utilized. Numbers are used. Data is collected. Graphs may be constructed. Data interpretations are made. And, some summary statement or conclusions may be reached.

Each sample hypothesis previously provided is an invitation to investigate. In experimentation good questions lead to good responses. The questions should be directed to the object under investigation, in this case, the seed. If you ask the right question, the seed will tell you what you want to know. We don't always get answers or totally correct answers to all our questions; but good questions and good search behaviors may bring us closer to a solution to our stated problem.

If you asked the seed:

Do you prefer rain water to tap water?

Will you grow to be the tallest plant if you initially have the greatest mass?

Will you grow better if I plant you deeper in the soil?

And, so on. The plant in response to the treatment will tell you. By your observations you will determine if your treatment resulted in a positive or negative reaction. If you ask the right questions, perform the right actions, and correctly interpret the responses, the seeds will have communicated to you, and possibly will have led you closer to a solution to your problem.

Experimentation is a powerful tool. Concomitant with the resolution of a problem is the self realization that one can solve or work towards the solution of problems. This act strengthens one's self concept for engaging future problems in and out of science.

o   o   o

IN SCIENCE, EXPERIMENTING IS NOT EVERYTHING, IT IS THE ONLY THING.

---

o   o   o

ACTIVITIES TO ENHANCE THE PROCESS SKILLS OF OBSERVING, INFERRING, AND PREDICTING

All of science begins with observation. From observation, inferences and predictions are made. Observations are communicated qualitatively and, when applicable, quantitatively. Quantification necessitates measurement and the assignment of a numerical value to that which is observed.

An observation is that which is communicated to us directly from the use of our senses. We see, hear, smell, taste, and feel something. This something is described by a set of observations.

Promoting the Process of Observation

Select any object. This could be a leaf, a rock, a button, a coin, a paper clip, a block of wood, a breakfast cereal with a unique shape, texture, etc., or a peanut. The object itself does not matter. However, to aid instruction, inherent to the object selected should be some intrinsic properties that are obvious and result in an extended dialogue of descriptions.

A loaf of raisin bread is a good example. Students observe the raisin loaf. The students describe the loaf as to color, shape, size, mass, texture, smell, any inclusions or protrusions visible on and in the exterior crust of the loaf, general configuration, taste (while the processes of taste and smell are approached with caution, this should be a harmless inclusion providing sanitary conditions are observed), and any associated temperature of the loaf. Observations can be expanded upon as students become increasingly proficient in the skill of observing. Over time, many of these descriptive observations will become routine in the consideration of other objects.

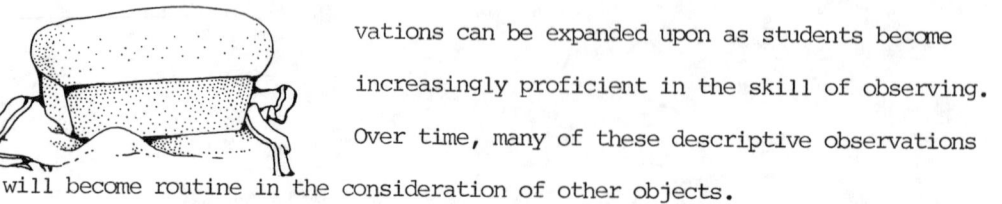

Promoting the Process of Inferring

The raisin loaf can be used as an example of an inference. An inference is an interpretation or explanation derived indirectly from the use of our senses. Again, observe the raisin loaf. From external direct views, one might see raisins partially in and out of the crust. From this, one might infer that there are raisins within the loaf hidden below the crust. One cannot be sure of this because of the lack of a direct observation of the inside of the raisin loaf. If one wished to establish the fact that this was true, the loaf of bread would have to be cut into and examined for raisins. This changes the inference to an observation.

Promoting the Process of Prediction

Predicting is one of the most important functions of science. A <u>prediction</u> is a forecast based on a combination of observations and inferences associated with related events. A prediction can be thought of as a refined inference. It is a resolution based on accumulated data. The data can take many forms. The more data that supports a predictive statement, the more accurate the prediction is assumed to be. The more times an action is observed with similar results the more validity is attached to the emergent prediction.

The raisin loaf example can be used to explain a prediction. One establishes the fact that loaves of bread with raisins partially in and out of the crust contain raisins within them. This has been established by slicing into the loaf. The repetition of this act, with confirming results, makes any future prediction about the containment of raisins within the loaf more valid. Confirmation of results with 5000 raisin loaves is more valid than confirmation with 500 loaves; 500 confirmations are more valid than 50 confirmations; and 50 confirmations are more valid than 5 confirmations.

Count the raisins in a slice of raisin bread. This is a quantifiable observation. Make an inference about the remainder of the raisin loaf. Make a prediction about raisin loaves.

SUGGESTED INSTRUCTIONAL ACTIVITIES FOR OBSERVING

🔺 Rubbing alcohol is contained in an unmarked bottle to which a few drops of food coloring has been added. This liquid is referred to as the "unknown." Place a drop or two of the unknown into one of each student's palm. Ask the students to observe what happens. Discourage the use of taste.* Compile, from the group, as many observations as possible. Ask...

> What happened to the unknown liquid?
> Was it absorbed or did it evaporate? The unknown liquid is alcohol and alcohol evaporates.
> Does it have an odor? Describe the odor. The description of an odor is a relative thing. Odor descriptions seem to be in the nose of the beholder. Odors are difficult to describe unless one compares them to a known identified odor. Hence, students say, "It smells like..." Most students describe the odor of the unknown as medicinal.
> Does it heat up, cool down, or leave the skin's temperature unchanged? Evaporation is a cooling process.
> How does the unknown liquid flow? What other liquids that you know flow like the unknown? Compare the viscosity of the unknown to that of water.
> What effect does the unknown liquid have on the surface of your skin? Any visible color change? Any cleansing effect to the skin? Did it remove any oil from your skin?

*   Advise the students not to taste this liquid.

*   Most students will state that the liquid disappeared. Rarely will they say, "Two drops of the unknown liquid disappeared in nine seconds." When applicable, students should include in their observations some reference to the time factor for the length of an occurrence. This is referred to as a space/time relationship and should become part of a full description.

🔺 Use a small, babyfood container. Fill this one-half to three-fourths full with cooking oil (olive oil works best). Place a small ice cube in the cooking oil. Ice, being less dense than oil, will float. The ice will melt changing into water. Water is more dense than oil and the water droplets clinging to the ice will be released and sink to the bottom in water spheres. Ask...

    What do you observe?
    What explanation can be given for the ice floating on top of the oil?
    What explanation can be given for the melted ice sinking to the bottom
        of the container?

*   Have the students make an inference prior to the action of placing the ice in the cooking oil. When it melts it contracts and its volume decreases. Does its mass also change?

△ Everyday relocate a common article in your classroom for example, move the wax begonia plant to a new location. Ask...

    What is different in the room?
    What has been moved?
    Where is it now?
    Where was it located yesterday?

*   Invite responses to the question, "What was I wearing yesterday? What was I wearing Monday of last week? What would you infer my favorite color is?" Each day should be punctuated with numerous, short observational challenges. If the principal, the nurse, or the janitor walked in and out of your classroom, turn this into an observational involvement. Ask, "What was this individual wearing? How tall was this individual? Did you observe any unique thing about this individual? Later, have any one of these individuals stand outside of the room and project one hand and part of their arm into the room. From this partial view of the individual, ask the students to infer which individual (the principal, the nurse, or the janitor) is standing out in the hall.

Instructor's description. Be careful with this one! Know your class or you may be unhappy with the responses. Ask...

How tall am I?
How much do I weigh?
Do I wear rings? If so, how many and on what fingers?
Do I regularly wear a wrist watch? If so, on what wrist?
How old am I?
With a short phrase or one sentence, describe me.
Infer the kind of car I drive.
Infer what hobbies I enjoy.
Predict....

* This list can be added to or subtracted from depending on you and your classroom situation. This involvement moves readily from direct observations and inferences to comparative observations and inferences. Ask the same questions describing the attributes of a different individual. This, and the prior information, can be recorded in two columns. Compare the responses to the questions for each individual. For example, is the other individual taller than you? Shorter than you? Or, the same height as you? When comparisons are made, the obvious question is "by how much do they differ?" This requires the additional use of the quantification component of observation.

More precise information can be added to observations by the inclusion of standard units of measure. When appropriate, use them. Standard units of measure are associated with particular devices to obtain measures. The following statements are examples of quantification and the devices used to obtain these measures.

I weigh 175 lbs. He weighs 195 lbs.
    The difference by comparison is 20 lbs. (weight scale)
Its mass was 78 grams. Its mass now is 73 grams.

The difference by comparison is 5 grams, five grams less. (balance)
The temperature is 22°C. The temperature was 17°C.
The difference by comparison is 5°C, five degrees more. (thermometer)

Quantitative observations are not always made with instruments such as graduated cylinders, rulers, balances, spring scales, etc. Sometimes expressions of approximate measures are used rather than precise values. These also are considered quantitative observations.

⚠️ Fill one clean glass with tap water. Fill another glass with an equal amount of tap water. To the second glass add one-half teaspoon of table salt and stir this well. Ask the students to observe these two containers and state their observations. How are these two liquids alike? How are they different? Have them use all their senses. When testing the water, use individual straws.

⚠️ In the automotive section of your hardware store, purchase one of the oil additives for car engines. Make sure that the material of the container is transparent and plastic. These oils usually have a much heavier viscosity than water. Invert the container. A bubble will rise to the top. Ask the students to describe this action. Ask them to make a quantitative observation, and to state an inference. Some interesting things are seen within the bubble as it travels upward.

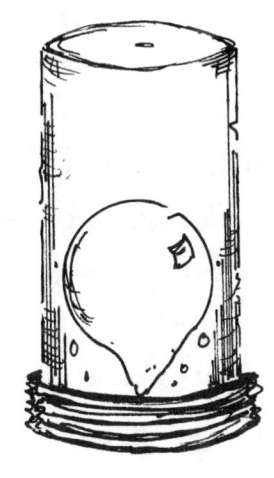

▲ Half fill a container with water. Observe the water. Is this a static or dynamic observation? Place several drops of olive oil or salad oil on the surface of the water and observe what happens. Is this a static or dynamic observation? Slowly add rubbing alcohol by pouring it carefully down the sides of the glass. What do you observe? Is this a static or dynamic observation?

* The water in isolation was in equilibrium with itself. It occupied the container and, other than evaporating slowly, it exhibited a static behavior. Oil was added. Its density is less than that of water. The oil adjusted itself to the environment and floated on top of the water. It can be observed depressed slightly into the surface skin of the water. Alcohol will mix with water, but added slowly it will float across the surface of the water. At this point the oil is no longer in contact with water, it is in contact with a surface layer of alcohol. This required an adjustment. A reaction or a moment of disequilibrium, wherein the oil dynamically adjusted itself to a new environment, was observed. Once adjusted, the oil settled down to a static existence. Thus, the observations have moved from static to dynamic back to static. Or, we can describe this in terms of before, during, and after observations each describing movement from equilibrium to disequilibrium back to a state of equilibrium.

▲ Invite the students to observe any object. Ask them to write down, in list form, all their observations. Instruct them to write down along side of each observation which one of the senses they used. When they are finished, ask them to tally up the individual senses to determine which sense they used most often. The least often. Point out that some senses are under used and some are over used. The more senses used, the more complete the observation is.

Sprinkle water on a piece of wax paper. Using the tip of a toothpick break the water droplets into smaller and smaller droplets. Ask...

Why doesn't the water stick to the wax paper?
* The water molecules' attraction for each other is greater than the attraction for the wax molecules in the wax paper. The attractive force of like molecules for each other is known as the cohesive force. Water molecules may also be attracted to molecules other than water molecules. This attraction of unlike molecules for each other is known as the adhesive force.

Why are drops of water circular?
* The cohesive quality of the water is continually pulling in the center of the water drop forming a sphere. This is the most conservative shape in nature. Geometrically, it is the most compact shape with the smallest possible exposed surface area.

As stated previously, observations can be described as static or dynamic. Some objects change so slowly that while they are not in a static state, they appear to be. The daily growth of a child is almost undetectable over a short, time span. Other objects respond more dynamically to change and new observations are obviously discernible. Events follow a pattern of being static and due to some chemical or physical change are set into motion. This motion generates a new, dynamic set of observations. When the dynamic action ceases, there is a resolution back to a static state. A set of observations for this condition would not necessarily match that of the original static observations. An example can describe this. Imagine an antacid tablet. This object can be fully described. This set of descriptions describes the tablet in a static state. If the tablet were dropped into a glass of water, a new set of observations would be engendered. This set of observations would describe the tablet in a dynamic state. The tablet would cease to effervesce. The process would have ended and, we have a return to a new static state requiring a new set of observations to describe this situation. Thus the pattern has repeated itself. Trained observers are aware that when a reaction is involved in an observation, three sets of descriptions need to be considered - static, dynamic, and a return to static.

🔺 Grow a morning glory plant. Observe how the tip of the stem moves in a circle. Record your observations. In which direction does it wind...clockwise or counterclockwise? How long does it take to make one full revolution?

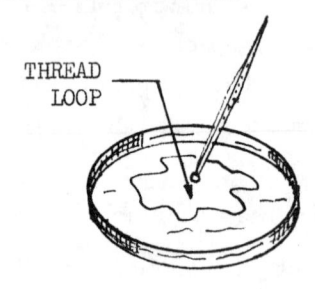

🔺 Make a loop from a 5 inch piece of silk thread. Place the loop in water. Touch the inside of the loop with a soapy toothpick end. Describe your observations.

    *   The thread is pulled into a circle.

    The addition of the soap to the inner circle of the thread loop weakens the surface tension inside the loop and the thread is pulled into a circle.

🔺 Sprinkle fine talcum powder on water in a shallow dish. Touch a soapy toothpick end to the middle of the floating powder. Describe your observations.

*   The surface tension at the point of contact of the soapy toothpick with the water is

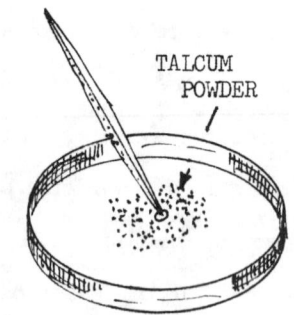

weakened and the powdered water suddenly appears to rupture as it is pulled in all directions by the stronger surface tension in the remaining areas of the dish.

△ Place one drop of airplane cement (or "Duco cement") in a shallow dish of water. Describe your observations. Include a space/time observation.

* Caution: most airplane cements contain acetone and butyl acetate. Thus, one must avoid breathing concentrated vapors. Also keep it away from heat, sparks, and open flames. In case of eye contact, immediately flush with water and call a physician. This is an excellent, high interest activity. Handled, noting the necessary precautions, by the teacher in the form of a demonstration wherein the teacher controls the airplane cement dispensing, etc., is recommended. This activity can be further enhanced by using a plastic shoe box with sufficient water to cover the bottom to a depth of 2 cm. Place a drop of airplane cement on the end of a wooden matchstick. Place this in the water and describe your observations. Good comparative observations can be made comparing this action to the prior drop-only activity. How can this action be explained? The action does not last forever. It does cease. Something is supplying the energy to the system. What is it?

△ Place a few drops of vinegar on a piece of chalk. Describe your observations in three stages, before, during, and after.

* Chalk is made from limestone. Limestone will effervesce in the presence of

acids, in particular dilute hydrochloric acid. Vinegar is a dilute form of acetic acid. A piece of limestone or a sample of baking powder will react in the same manner.

🔺 Fill a milk bottle or some other suitable glass container with hot water (not boiling). Invert the bottle over a clear, plastic shoebox. After the water spills out lower the neck of the bottle into the water. Describe your observations. Explain what you see.

*   As the air inside the bottle cools, it contracts and the water level inside the bottle rises.

🔺 Place a short candle inside a glass container. Light the candle. Slowly pour soda water (carbonated) into the container surrounding the candle. Describe your observations.

*   Carbon dioxide from the soda water combines with the carbon dioxide given off by the candle. Carbon dioxide does not support combustion. Carbon dioxide is heavier than the oxygen needed to support combustion. The rising level of carbon dioxide eventually surrounds the candle and the flame. Since no oxygen can get to the burning candle, it is snuffed out.

🔺 Obtain a dried prune. Describe it. Place the prune in water for a twenty-four hour period of time. Describe it again. How do the two sets of observations compare?

🔺 Observe the lines contained within the palm of your left hand. Compare these observations to those lines contained within the palm of your right hand. Are they alike?

▲ Observing the paper worm. Using an individually, wrapped soda straw open one end of the straw wrapper. With the soda straw in an upright position, open end at the top, gently tap the straw on a flat surface working the straw wrapper carefully down to the bottom of the straw crimping it along the way until you have a tightly, crimped wrapper about 1½" to 2" long. Slide this off the straw retaining the tight, worm-like crimp. Using an eye dropper, add several drops of water down the length of the crimped wrapper.

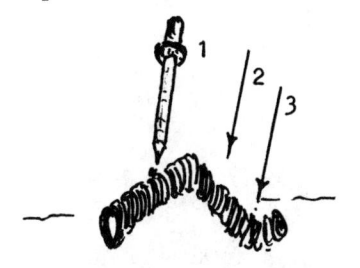

*   The fibrous material of the wrapper absorbs the water causing the worm to expand. The worm appears to be alive and moving. This movement can be shown clearly and dramatically using an overhead projector. In some localities individually wrapped soda straws are difficult to get. Most fast food establishments are most cooperative in donating these soda straws for use in your class.

▲ Does a leaf grow equally in all directions? Superimpose a finely-meshed screen (window screening works well) over one leaf of a plant. With the screen in this position, gently rub an indelible, black-ink marker across the screen so as to leave a pattern of equally-spaced ink dots on the leaf caused by the ink reaching through the wire mesh. You have now established a uniformly marked grid. Allow the leaf to grow for a period of time. Periodically, observe the spaces between the previously, established uniformly spaced dots. Have they changed? Does a leaf grow uniformly in all directions? Do certain parts of the leaf grow faster than others? Would your observations be consistent for the growth of other parts of the plants, for example, the stem or stalk? Do leaves of all plants grow in the same manner and rate?

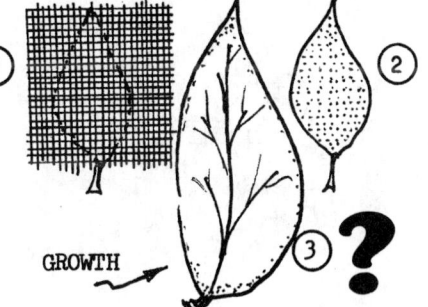

SUGGESTED INSTRUCTIONAL ACTIVITIES FOR INFERRING

△ A

△ B

△ Examine drawing A and B. Make an observation of each drawing. Make an inference as to what is going on in each situation.

Examine drawing A and B. How are they alike? How are they different? Make an inference for drawing A. Drawing B. Make a prediction for drawing A. Drawing B.

△ A

△ B

54

🔺 Fill a plastic, sandwich bag about one-half full with water. Seal the bag. Insert, at any angle several, sharpened wooden pencils into, through, and out the other side of the plastic, sandwich bag. What do you observe? Make an inference about your observation.

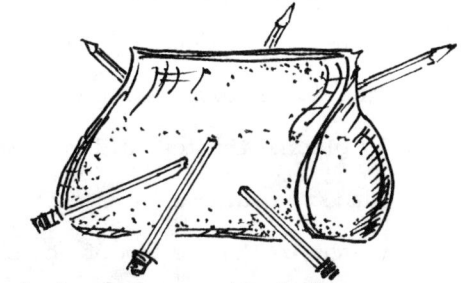

* The plastic closes immediately about the shank of the pencil at those places where it pierces the sandwich bag. This seals off any water leaks.

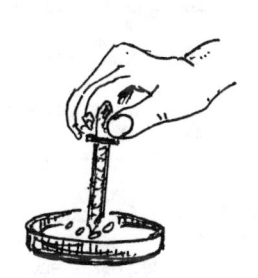

🔺 Place a dry penny on a flat surface. Using an eye dropper, slowly place water a drop at a time onto the penny until no more will fit onto the surface of the penny without running off. Make an inference as to whether or not the same number of drops will fit on the reverse side of the penny.

* The penny through the cohesive action of the water, will hold about 60 drops of water on its surface.

🔺 Place an ice cold, soda pop container in a warm room. Describe what happens. Make an inference.

* The container begins to collect moisture on the outside of the container. Inference - It leaked out of the container. The liquid is passing through the wall of the container. What happens is: the mass of the cold container placed in contact with the ambient temperature of the room causes water vapor to form and collect on the side of the container (condensation). The air in the vicinity of the container being cooled down cannot hold as much water vapor as the warm air. When the cold container warms up to room temperature the air which surrounds the container has a greater capacity to hold water vapor. It absorbs the previously formed water droplets back into the air (evaporation). This is the evaporation/condensation process. This can be dramatically shown by placing the ice cold container in one arm of an equal-arm balance. Quickly place the balance in balance. The condensation process will quickly add weight to the one arm of the balance and place the balance out of balance. An inference - Will the balance return to a point of balance as the evaporation process takes over?

🔺 Inferring can be promoted in a variety of ways. Ask the students to close their eyes and place their heads down on their desk. Ask them to remain this way until signaled to do otherwise. In this position, have pre-arranged actions staged by one or more members of the class. These actions can take many directions, for example, one student can be positioned in each corner of the room. In some pre-determined sequence these students perform some acts utilizing any one of the senses except sight. One student can make a noise such as a low whistle; another one can emit a spray from a container containing some scented odor; another one can wind up some apparatus to convey the clicking of gears; and the remaining one, in stocking feet, can walk about the room dragging a piece of string across the heads of those students in their seats with their heads down. All these involvements could be embellished or reversed. The seated students are invited to lift their heads up and contribute to the reconstruction of the events and, where applicable, make suitable inferences.

🔺 Obtain a large, cardboard packing crate. Cut out the top or one side of container. Rest the container on the floor. Invite one or more volunteers to crawl inside the box. Have students outside the container perform actions to communicate all the senses except the sense of sight. The volunteers are to infer what is going on outside the container. Some of these events may need to be rehearsed to convey unique actions.

▲ Cover a young plant such as a mung bean plant or a young radish plant with a paper cup that has a hole cut out its side. Infer the direction of growth the plant will take under these conditions. Observe and compare these observations to your previous inference.

*   The plant should seek the light and grow out of the hole and upward.

▲ Place your hand above and below a container of ice cubes. What do you observe? What can you infer from these observations?

*   You should observe that your hand feels colder under the container as opposed to placing it above the container. You might infer from this that cold air is heavier than warmer air and sinks to the lower areas being replaced by warmer, lighter air in the higher elevation areas.

▲   MAKE AN INFERENCE.

SUGGESTED INSTRUCTIONAL ACTIVITIES FOR PREDICTING

Predictions become more valuable when they become more accurate. Calculated predictions are not always one-hundred percent correct. No prediction is. Repetitive trials, accumulated, well collected data, astute observations, and careful analysis of patterns, trends, recurring cyclic events all reduce the chance of incorrect predicting. In the act of predicting, it is important to:

Carefully collect relevant data consistent with the observations.

Identify relevant variables; concentrate on those variables that have a bearing on the results.

Search out the cause-effect relationships. What impinges on what? What caused the event to occur? And, what was the result or effect?

Challenge yourself to seek similar results with a higher level of consistency. Are the results reliable?

Summarize your results. Use opportunities to forecast (predict) future observations based on past efforts and data.

Expect the unexpected at all times. Look for relationships that link one thing to another.

Constantly look for repetitive cycles, events, patterns, or models which provide clues for future predictions based on established rhythmic occurences.

▲ A rubber ball is dropped from a height of three feet. It is observed that upon striking the floor it rebounded to a height of twenty-four inches. Repeating this act, one could infer that when the ball is dropped again from the same height, it will rebound to approximately the same height. The ball is dropped one-hundred times from the same height and the various heights of each rebound are recorded. From this accumulated data, one could predict, the rebounding height of the rubber ball when dropped from the same height. Find a ball. It can be a golf ball, a rubber ball, a ping-pong ball, or a super ball. Drop it from a known height. A desk top's height will do. Drop the ball by releasing it in the same manner everytime. Record the height of the rebound. Do this in every case. Find the rebound average. What do you predict the rebound height will be if you dropped the ball from the same height one more time? Twice the height?

I <u>predict</u> that I will graduate from high school.

△ Some facts are:

> I.....am 14 years of age.
> am a male.
> am left handed.
> wear glasses.
> am a smoker.
> do get good grades in those subjects
>   I enjoy.
> hate taking English.
> enjoy math and science.
> watch TV at least 3 hours a day.
> have many hobbies.
> like to eat, but I am not overweight.
> skip a lot of my classes. If I don't
>   feel like going to class, I don't.
> manage to con people quite easily.
> earn my own spare money.
> own two, new bicycles.
> worry a lot.
> am an introvert.
> don't like intellectual engagements,
>   but I can apply myself when
>   scared.
> lack perserverance.

Examine the facts. Establish what are the important facts relative to the stated prediction. Analyzing these facts, do you agree with the prediction?

△ A golf ball is rolled down an inclined plane. The inclination of the plane is twenty degrees. The golf ball is rolled down once, then seventy-five times more. The first time the ball was released it rolled down and off the ramp a distance of forty-eight inches from the ramp's end. The average of the next seventy-five rolls was 47.9 inches. State a prediction for the next roll down the inclined plane. If the inclined ramp were to be raised an additional ten degrees higher, what might you infer about any changes in the distance the same golf ball would roll out from the ramp's end? What prediction, if any, can you make about the change in the height and the consistency of the distance of the roll?

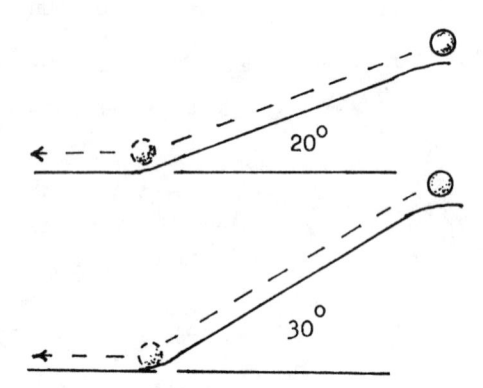

△ Place one of your hands flat on a sheet of paper. Using a pencil, trace the outline of your hand. Observe the hand trace. Which finger is the longest? The shortest? Measure these fingers (quantitative observations). Using this information, infer which fingers are the longest and the shortest for other members of your class. Have the students make the necessary measurements. Collect this data. Compute the average length for the longest finger and the average length for the shortest finger. What predictions can you make for similar students in other comparable grade levels?

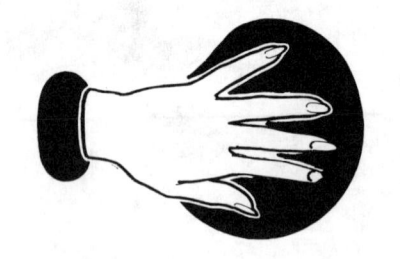

50 mL 50 mL

A B C
40°F 100°F ?°F

🔺 If two glasses of liquid marked A and B were poured into glass C, what would the temperature of this mixture be?

* The resulting mixture would be 70°F. This is the arithmetic average of the temperatures of the two liquids.

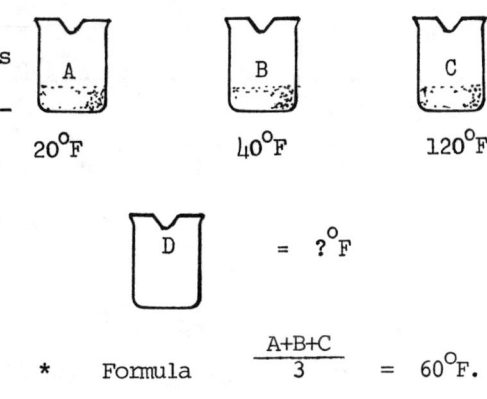

🔺 What will be the temperature of the liquid in glass D, if the liquids in glasses A, B, and C (with temperatures of 20°F, 40°F, and 120°F) were all poured into glass D? Glasses A, B, and C have equal quantities of liquid.

A B C
20°F 40°F 120°F

D = ?°F

* Formula $\dfrac{A+B+C}{3}$ = 60°F.

🔺 Fabricate your own example and calculate the final temperature of the mixture. Verify this by using actual materials. Discover the rule for predicting the temperature. Would you infer that the rule changes if unequal quantities of the liquids of varying temperatures are mixed as opposed to uniform quantities of the liquid being mixed? Devise a technique to establish evidence for your rule.

⚠️ Make a prediction. What will happen to the price of a stamp for a 1-ounce first-class letter in the next five years?

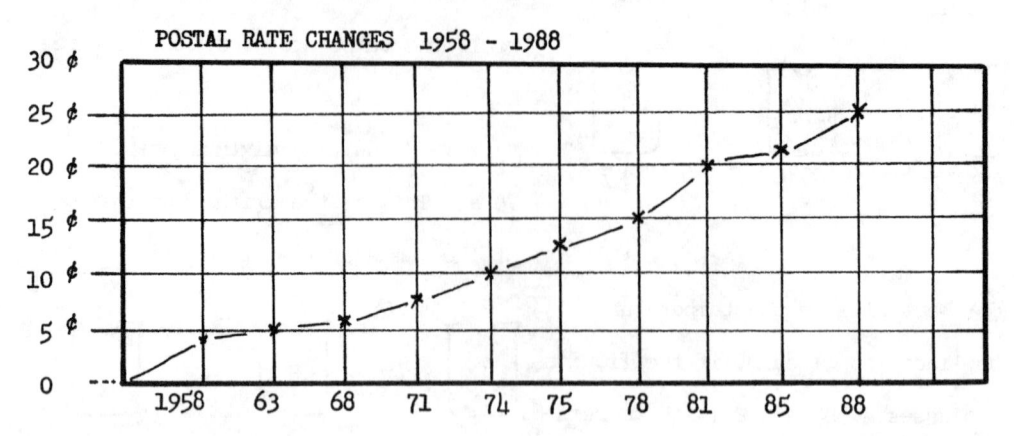

POSTAL RATE CHANGES 1958 - 1988

(Graph showing postal rates from 1958 to 1988, with prices rising from under 5¢ in 1958 to about 25¢ in 1988, across years 1958, 63, 68, 71, 74, 75, 78, 81, 85, 88)

* The data shows that for the last thirty years, the price has increased. One could predict that this trend would continue.

⚠️ Make a prediction. A student was asked to do as many push-ups as possible in a one-minute period, resting for two minutes, and then to do as many push-ups as possible for another one-minute period. Then, again a two-minute, rest period. And to continue this for a period of five, one-minute attempts.

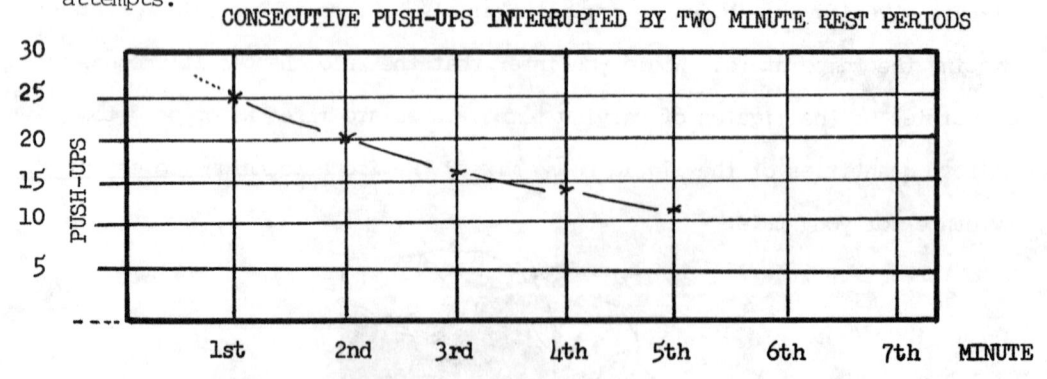

CONSECUTIVE PUSH-UPS INTERRUPTED BY TWO MINUTE REST PERIODS

(Graph showing push-ups decreasing from about 25 in the 1st minute to about 12 in the 5th minute, with x-axis labeled 1st, 2nd, 3rd, 4th, 5th, 6th, 7th MINUTE and y-axis labeled PUSH-UPS from 5 to 30)

* An observation of the data reveals that the student is tiring and it can be predicted that the number of push-ups will lessen as the one-minute performances continue on.

SUGGESTED INSTRUCTIONAL ACTIVITIES FOR CLASSIFYING

Classification is the process of organizing objects based on similarities and differences. It is something we all do. It is natural and efficient. We are natural born classifiers. We classify ourselves, our students, our belongings, and objects of the universe. We were born classified. At birth we were carefully examined. Our weight, length, footprint, any birth marks, and other pertinent observations were recorded. These observations are called characteristics, attributes, or traits. They describe us individually and collectively. They distinguish us one from another.

There are two methods of classification, serial and binary ordering. Serial refers to the arrangement of objects in series, rank, or row according to some numerical attribute. Examples are: arranging objects in descending order from large to small, from heavy to light, of from high intelligence to low intelligence. These same attributes could have been serially arranged in a reversed (ascending) order. Binary means "two." When one classifies objects using a binary system one divides or sorts the objects into two piles or groups. One object possesses the attribute being considered. The other does not. The negative connotation is an all encompassing one. This has its advantage. The counterpart of, "It is red (positive statement)." is "It is not (negative statement) red." This negative statement says that the object or objects can be any color but red. This is an all inclusive, efficient statement which facilitates the construction of classification systems. A serial classification system can be transformed into a binary classification system by the establishment of a selected criteria. For example, if the attribute being observed is age and serially the ages have been arranged from 16 years of age through 32 years of age, two groups can be created; one group under twenty years of age; and, the other group being not under twenty years of age.

Schematically, a classification map for seven items would look like this:

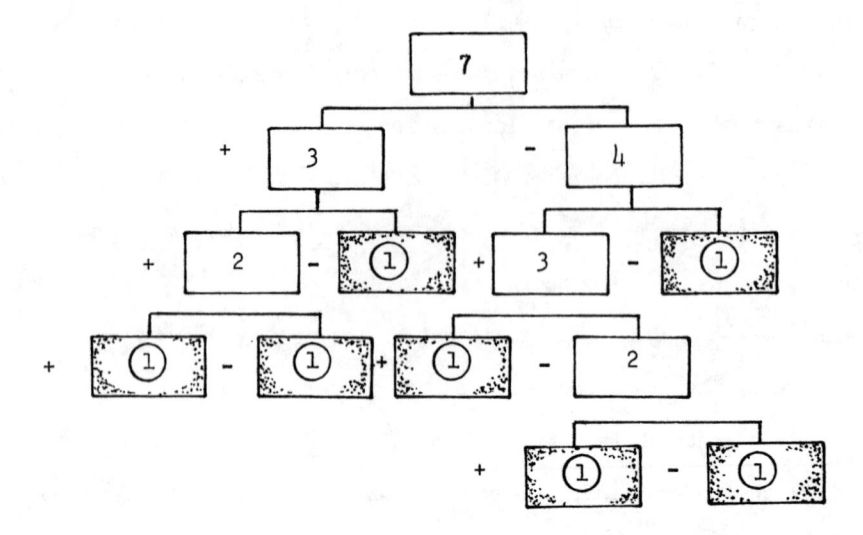

A binary multi-stage classification system is said to be complete when the number of single-description cells located at the bottom of the flow diagram equal the orignal number of objects contained in the initial cell.

⚠ Set up a classification map for the following objects

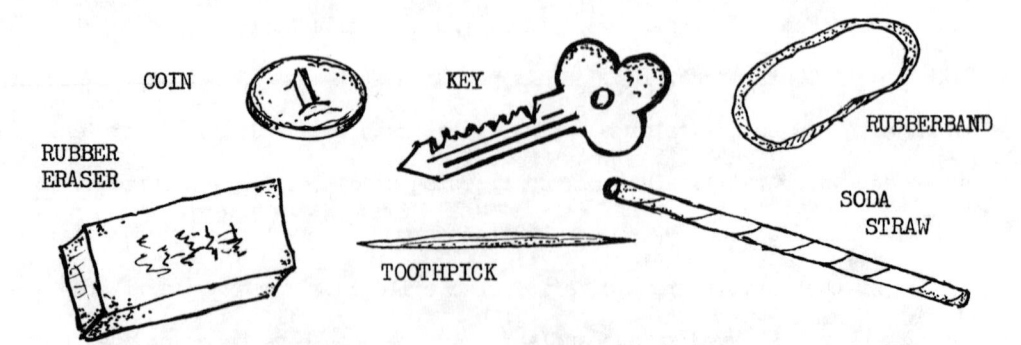

COIN KEY RUBBERBAND
RUBBER ERASER TOOTHPICK SODA STRAW

This activity will be best served if you can obtain the "real" objects rather than attempting to observe from these sketches.

* Count the number of items in the set of objects being observed. There are six items. This number should be written in the first cell of the evolving classification map. The final number of single-description cells should also equal this number. Write a complete set of descriptions (attributes) for each item, for example, the rubberband is elastic, brown in color, cylindrical in shape, etc. Examine each array of attributes describing each item in the set. Look for common attributes to link two or more objects into groups based on similarities and differences. For example, the key and the coin are both metallic. When preparing to divide objects into two groups, expressed positively and negatively, look for attributes that will allow, as nearly as possible, the same number of items to be grouped in respective cells. This works towards efficiency in the classification system. To have one object expressed positively and the remainder of the set of objects expressed negatively and to continue with this pairing is inefficient. Winnow out the attributes that provide you with a good balance of numbers of objects in corresponding cells.

Think about it.

> Is there only one correct way to group objects in a classification system? Explain.
>
> Is any one way to classify a set of objects better than another way?
>
> Is your educational system based on a classification system? Explain.
>
> Did you ever attempt to classify your friends?

▲ Observe a collection of rocks. Do not concern yourself with naming the rocks. Naming is not parcel to classification. Properties, attributes, and traits are. Set up a classification system based on your observation.

🔺 Observe a collection of leaves. Set up a classification system based on your observations.

🔺 Observe a collection of buttons. Set up a classification system based on your observations.

🔺 One of the purposes of a classification system is to provide for the identification of objects. Observe the Scoofors. What attributes must an object have to be labeled a Scoofor? Write a brief description for a Scoofor.

These are Scoofors.

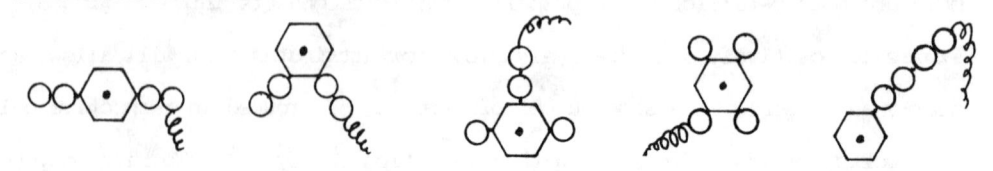

These are not Scoofors. Why are they not Scoofors? Examine each one and list what necessary attributes are lacking for each to have it qualify as a Scoofor.

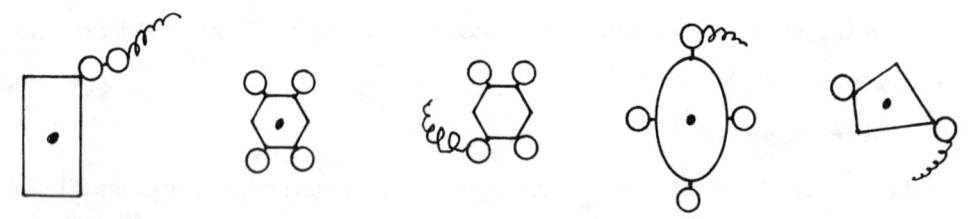

Which of these are Scoofors? Which ones are not Scoofors? State what attributes the non-Scoofors lack in order to qualify as Scoofors.

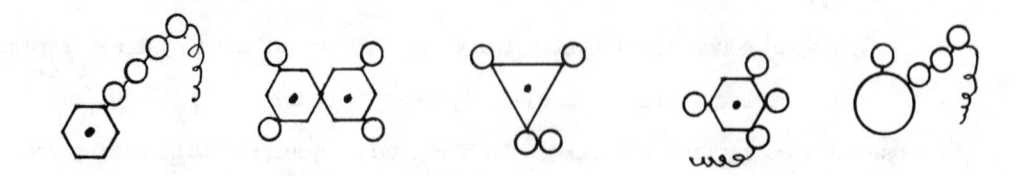

Often times more can be learned by constructing examples than by using someone else's example. Construct your own Scoofor-type observation example. One learns more by building a house than by buying a completed one. Try one! These are Knubblenoses. These are not Knubblenoses. Which of these are...?

Is there only one way to classify? Or, are there many ways with some preferred ways? Comparisons of various classifications systems should be made with the idea of selecting a preferred way to classify a specific set of objects. What would it be like if objects in the supermarket were not classified? Why are the groceries organized in a special way and then periodically changed?

▲ On one half of the bulletin board have the students' papers lined up alphabetically, horizontally, and all facing the same way. On the remaining half, have another set of the students' papers arranged randomly, upside down, backwards, stapled over one another, and so forth. Which is the preferred arrangement? Why? Does classification help us? How?

▲ The alphabet is a collection of 26 letters. Ask the students how they might classify the letters of the alphabet, for example, vowels, consonants, shape, configuration, etc.

* Ask...

>   How and what do libraries classify?
>   How is a newspaper organized?
>   How are stamps classified?
>   How are countries classified?
>   How are coins classified?
>   How are stocks classified?

▲ Collect a random assortment of leaves, breakfast cereals, or assorted size and color rubberbands. Set up classification systems for the **various** groups.

▲ Use assorted blocks of wood of different geometric shapes, sizes, colors, and textures (tinker toy parts work well here), allow for the physical, concrete organization of various classification systems.

▲ Invite the visually handicapped to classify through the sense of sound, smell, and touch.

▲ On special holidays students may have gone from house to house repeating, "trick or treat?" They usually end the day with a bag of assorted treats well intermixed. Any one item is difficult to find. How might they solve the problem of readily locating specific treats from their collection?

▲ Go outside to any bicycle rack (the more bicycles, the better). Observe the bicycles.

> How many bicycles are there?
>
> What is the most popular color? The second most popular?
>
> How many different manufacturer's brands are there? Which brand name is the most popular?
>
> How many and what variety of different (gear ratios) bicycles are there? Which gear ratio is the most popular? How many are there of each?
>
> How many bicycles are locked?
>
> Approximately how much money (total dollar value) do these bicycles represent?

Your observations furnished you with data. Data can be useful if it helps us solve some problem or dilemma.

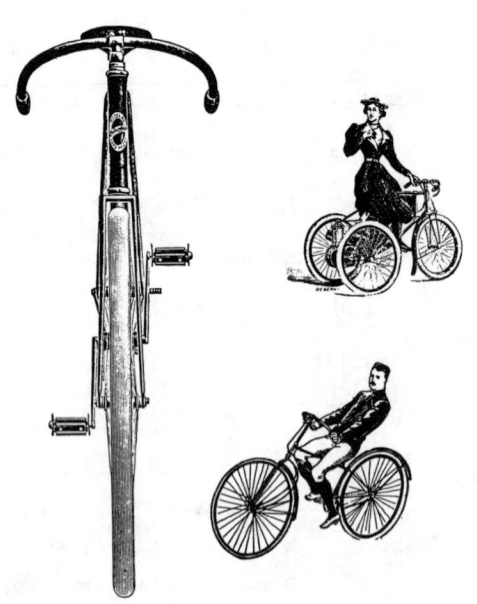

Based on your observations, if you were a bicycle manufacturer and you wanted to make use of this information (data), what would you do to assure that your observations are not false or invalid? What information would lead you to consider changes in the product you now manufacture? What might you predict the most popular bicycle style, color, gear ratio, etc. would be if you checked numerous other bicycle racks? If you were a bicycle manufacturer, what color, size, style, etc. would you expect to be your best seller?

▲ Classify the following items.

A

B

C GARLIC

D

E THYME

F

G

⚠ Using keysort cards for classification.

The binary system of classification lends itself very well to keysort cards for classification. The binary system can be likened to a light switch. Either the electricity is on or it is off. And so it is with keysort card criteria statements. They are either positive or negative, "It is... or, It is not..." Any 3 x 5 index card or Manila folder cut to a suitable size will do. Using a hole punch, duplicate the model. You may wish to change the criteria statements to suit your needs.

INDIVIDUAL		INDIVIDUAL JOHN DOE	
⊙ Under 15 years of age	Over 15 years ⊙ of age	⊙ Under 15 years of age	Over 15 years ⊙ of age
⊙ Is a senior	Is not a ⊙ senior	⊙ Is a senior	Is not a ⊙ senior
⊙ Speaks Russian	Does not speak ⊙ Russian	⊙ Speaks Russian	Does not speak ⊙ Russian
⊙ Drives a car	Does not drive ⊙ a car	⊙ Drives a car	Does not drive ⊙ a car
⊙ Has completed a driver education course	Has not completed a driver education course ⊙	⊙ Has completed a driver education course	Has not completed a driver education course ⊙
⊙ Can swim	Cannot swim ⊙	⊙ Can swim	Cannot swim ⊙
⊙ Overweight for body size	Not overweight ⊙ for body size	⊙ Overweight for body size	Not overweight ⊙ for body size

UNPROCESSED CARD   PROCESSED CARD

*   Positive or negative cutouts are made for each response for each statement on the card from each individual in the class. The cutout reaches into the punched hole making a long "U" configuration. The processed card has been cut into and forms a profile for one individual. As read from the card, this individual is over 15 years of age, a senior, speaks Russian, drives a car despite not having completed a driver education course, swims, and is not overweight for his body size.

To use the keysort cards to their maximum, make a complete deck of cards comprised of all the individuals in your class or group. All cards should be processed for responses. With the cards smoothly stacked, insert a finishing nail through all the cards at any one of the seven questions selecting either the positive or negative counterpart of the question. Shake the deck. Cards may fall out. Depending on the question selected and the responses to the question, all cards may be retained on the nail, none of them may be retained, or some of them retained. For example, if you wished the names of those individuals who are overweight for their body size, insert the nail, shake and observe. There may be no individuals overweight, hence no cards will fall from the deck. Or, one or more cards may fall. If you wish to know those individuals who are both overweight and do not swim, you must first probe the overweight question and from the cards that fall out (if any) form another, smaller deck and probe the "cannot swim" question. Any falling from this probe will be those individuals who are overweight and cannot swim. If desired, this type of inquiry can be carried further to compile an aggregate of individuals with discrete traits. This same technique can be used to form decks of cards describing rocks, trees, plants, etc.

SUGGESTED INSTRUCTIONAL ACTIVITIES FOR GRAPHING

A graph is a device for communicating a relationship between two variables. These variables are the manipulated (independent) variable and the responding variable (dependent). Properly drawn graphs contain a large amount of concise information. This information can also be useful in generating additional information not plotted on the graph. In an experiment all variables are held constant but one which is manipulated, hence the term "manipulated" variable. The reaction to this manipulation is some form of a response (or reaction), hence the term "responding" variable. An example would be a plant retained in darkness and one retained in light, and corresponding measurements of growth recorded. The light factor is manipulated and the response to this manipulation is the growth factor. A graph can record this action.

Graphs consist of scales on two axes that are usually equal. The independent variable or the manipulated variable is plotted on the horizontal or "X" axis and the dependent variable or the responding variable is plotted on the vertical or "Y" axis. Both axes usually start at zero.

Graphs can take many forms, but the rectangular coordinate form is perhaps the most common and useful. This system employs two intersecting lines perpendicular to each other. The horizontal line (X axis) and the vertical (Y axis) intersect at a point called the origin which is labeled "O." Negative values of "X" are located to the left of the origin, positive values are to the right. Negative values of "Y" are located below the origin, positive values are above the origin.

Components of a Graph

    Title - Make sure that the title communicates the purpose of the graph. Does it answer the question, "What is the graph about?"

Label the axes - Use the "Y" axis to plot the dependent or responding variable. Use the "X" axis to plot the independent variable or manipulated variable. Label these axes clearly describing the variables tested.

Find the range of values for each variable. Decide on scales for these values enabling the scales to fit within the confines of the graph paper.

Plot the data using easily understood symbols that are accurately plotted to show the relationship between the variables.

Draw a smooth, "best-fit" line or curve that goes through as many points as possible, avoiding a "waffling or zigzagging" path.

The curve of a graph can represent a direct proportion, inverse proportion, or linear relationship.

A qualitative definition of a direct proportion graph is: as one variable increases, the other increases proportionately; as one decreases, the other decreases proportionately. The relationship is direct. Direct proportions when graphed, trace a straight line passing through the origin. The inclination of the line can vary but it still must pass through the origin to qualify as a direct proportion graph.

DIRECT PROPORTION GRAPH

In an inverse proportion as one variable increases the other decreases proportionately. For example, if one variable doubles, the other becomes one-half. The shape of the line traces one-half of a hyperbola.

If a straight line does not pass

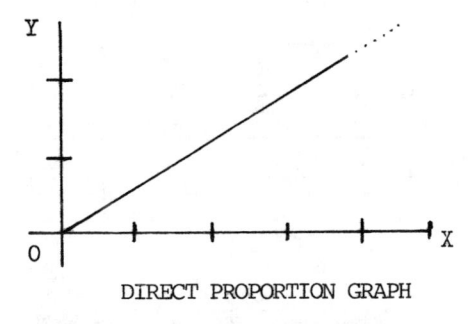

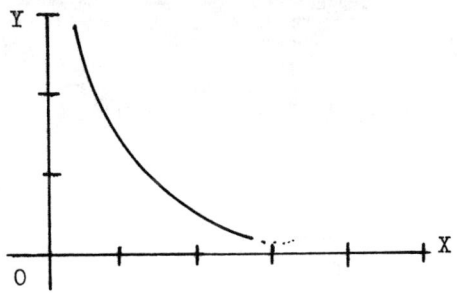

INVERSE PROPORTION GRAPH

through the origin, then the graph does not show a direct relationship. The relationship is linear but not a direct proportion.

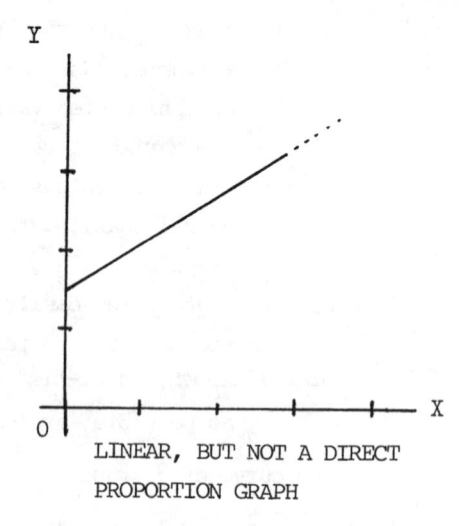

The construction and use of bar graphs precedes the use of line graphs in the elementary schools. A histogram is a bar graph that records the frequency of occurence of data collected in an experiment or activity. For example, a collection of the various heights of students in a class is gathered. The data is posted in bar graph form in terms of one column representing grouped data or similar heights as well as singular occuring, height data.

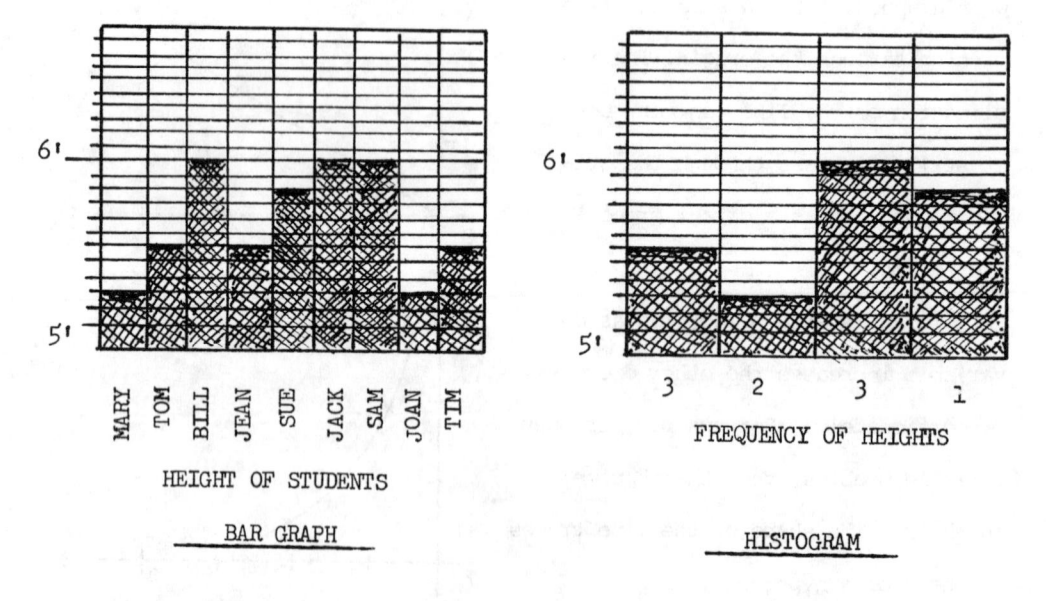

LINEAR, BUT NOT A DIRECT PROPORTION GRAPH

HEIGHT OF STUDENTS

BAR GRAPH

FREQUENCY OF HEIGHTS

HISTOGRAM

Can a line graph be constructed from a bar graph?

74

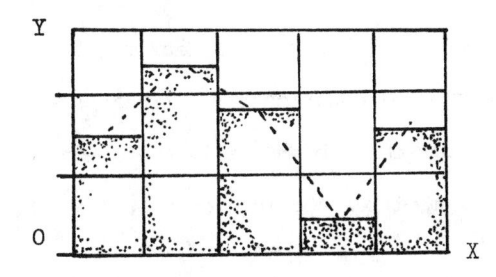

Observe the bar graph. Place a dot at the mid-point of the top of each vertical bar. Connect these dots. Voila, a line graph is born!

Abstracting additional information from graphs

Graphs communicate much valuable information. Much of this information is not directly plotted on the graph. It is inherent to the graph and can be abstracted through the process of interpolation and extrapolation. Interpolation is the process of predicting information that is derived from an examination of values within a measured range of previously plotted values. Extrapolation is the process of predicting information that lies beyond the range of observed and plotted values. Both opportunities for prediction are inferences. The validity of these inferences is predicated on the existence of highly reliable patterns of plotted values. Observations that are erratic, resulting in the plotting of such values into a erratic pattern, do not make for reliable interpolations or extrapolations. The further away one gets from observed values the less accurate are the predictions. Extrapolation is usually a less reliable form of predicting than interpolation. Whenever possible test your interpolations and extrapolations by observation.

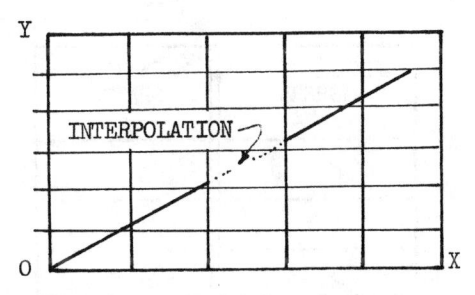

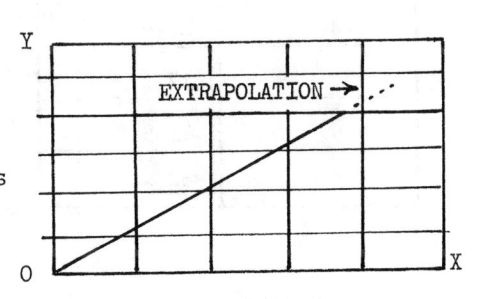

Determining the slope of the graph.

It is possible to obtain the equation representing the linear relationship of the variables of the graph. The general equation for a linear relationship is Y = MX + D. The slope is represented by M. X is the symbol for the X axis coordinate. And D is the symbol for the intercept of X and Y, when X is equal to zero.

When data are plotted in a linear relationship, the plotted points lie in a relatively straight line. A very useful function of a graph is slope. The slope of a linear graph is represented by the following equation:

$$\text{Slope} = \frac{Y^2 - Y^1}{X^2 - X^1}$$ This can also be expressed as: $$\text{Slope} = \frac{\text{Rise}}{\text{Run}}$$

When the slope of the linear graph is determined, you can, by using the following formula, determine the "Y" coordinate, Y = MX + D.

NUMBER OF WASHERS	MASS IN GRAMS
0	0
1	4.3
2	8.6
3	12.9
4	17.2
5	21.5
6	25.8

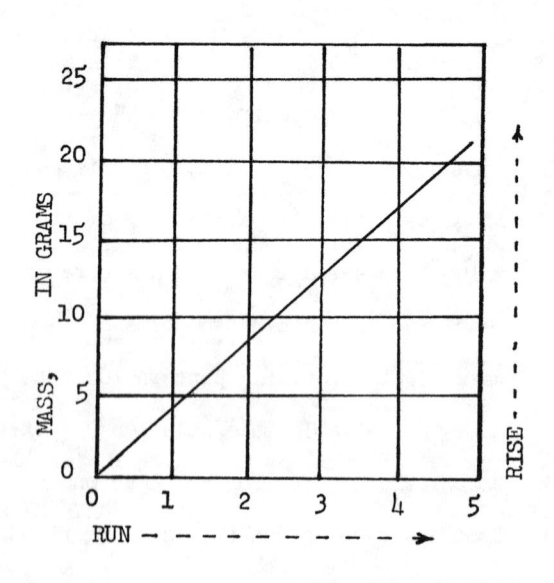

The slope is _____.

* 4.3.

The slope is _____.

The Y Coordinate is _____.

🔺 The following is a collection of data obtained from viewing a student who attempted to do as many push-ups as was possible for that individual in as short a period of time as possible. The student started off quickly and over a period of time slowed down. The number count per minute was recorded as follows:

first minute	35 push-ups
second minute	25 push-ups
third minute	16 push-ups
fourth minute	8 push-ups

* Construct a line graph of this action. What will your manipulated variable be? What will your responding variable be? From your graph and the plotted data, make an inference. Make a prediction.

🔺 The following is a collection of data obtained from a study of smoking and coronary artery disease as a cause of death. The data is as follows:

SMOKING STATUS	*RELATIVE DEATH RATE	(coronary artery disease)
Never smoked	1.0	
1 - 9 cigarettes smoked daily	1.2	
10-20 cigarettes smoked daily	1.6	
More than 20 cigarettes smoked daily	1.9	

* "Relative death rate" is the death rate among smokers divided by the death rate found among nonsmokers.

* Construct a line graph of this data. What will your manipulated variable be? What will your responding variable be? From your graph and the plotted data make an interpolation and an extrapolation. Which one of these approximates an inference? A prediction?

🔺 In response to the question, "How does the temperature of water effect its solubility?" three beakers were each filled with 100 mL of water. Each beaker's water temperature was raised to different temperatures: 20°C, 40°C, and 60°C respectively. At each temperature, sugar was dissolved in the water. The water in each beaker dissolving as much sugar as it was capable of before reaching a saturation point which allowed the excess sugar to precipitate out of solution. The water in the 20°C beaker dissolved 204 g of sugar; the 40°C beaker dissolved 238 g of sugar; and the 60°C beaker dissolved 287 g of sugar.

* Construct a graph of this data. What will your manipulated variable be? What will your responding variable be? What would you predict the amount of sugar would be that would be dissolved in 100 mL of water at 80°C?

> *** The amount of solute (sugar needed to saturate a solution) is a measure of the solution's solubility. Solubility is often expressed as the number of grams of solute that can dissolve in 100 g of solvent.

Two rabbits (male and female) were placed on an island previously devoid of rabbits. The rabbit population increased with time. The data collected on the rabbit population is as follows:

```
        Initial start........two rabbits
     six months later.......five rabbits
   twelve months later......twelve rabbits
 eighteen months later.....two-hundred rabbits
```

* Construct a line graph of this data. What will your manipulated variable be? What will your responding variable be? What prediction or predictions might you make about this situation?

Three lima beans were planted under identical conditions. A nutrient was administered to these plants. This was mixed with water in three different concentrations: 5%, 10%, and 15% per unit of volume. Each week, growth measures were recorded. These measures are as follows:

GROWTH RATES

	1st week	2nd week	3rd week	4th week	5th week
Plant receiving 5% solution	2 cm	6 cm	10 cm	14 cm	22 cm
Plant receiving 10% solution	2 cm	9 cm	14 cm	21 cm	32 cm
Plant receiving 15% solution	2 cm	7 cm	10 cm	12 cm	15 cm

*   Construct a line graph for each plant. Let the manipulated variable be time and responding variable growth. Compare one graph to another. This can be done through single graphs or all three growth patterns can be plotted on one graph using different notations or colors to distinguish one plant from the other. Which concentration of the nutrient provided for the best growth pattern? For each plant, which week interval provided the maximum growth? Make some inferences relative to the application of the various concentrations of the nutrient. What conclusion might you have arrived at? Design an experiment to further substantiate your inferences or conclusions.

A ball was dropped from several distances above the floor. The height of the rebounds and height of the drop were recorded as follows:

Height of the drop in feet	Height of the rebound in feet
2	.5
3	1.8
4	2.3
5	3.0
6	4.0
7	4.5

*   Construct a line graph of this data. What will your manipulated variable be? What will your responding variable be? What height would you predict the ball would rebound to if the height of the drop was increased to eight feet?

SUGGESTED INSTRUCTIONAL ACTIVITIES FOR IDENTIFYING AND CONTROLLING VARIABLES AND TESTING HYPOTHESES

If one knows what one's objectives are, and if one has a well-conceived strategy for reaching these objectives, one has a good chance to succeed in the fulfillment of these objectives. This is also the way of experimentation. If one is clear about what one is searching for and if the design of the experiment complements the search question or statement (the hypothesis), solutions come. Not always, nonetheless, continued action moves one closer to solutions.

Experimentation in science is the testing of hypotheses. To accomplish this, a scientist must identify those variables that will influence the outcome of the experiment. These variables must be controlled. These variables must remain constant or unchanged during the experiment. This is necessary so that the assignment of subsequent credit for an action can be attributed exclusively to that one variable which is manipulated and to no other variable. More than one manipulated variable confounds the experimenter for no accurate match of action/reaction of variables can be assigned to one specific variable. That variable which is germane to the stated hypothesis is manipulated. It is the manipulated variable that causes the experimental result which is viewed through the reaction of the responding variable.

A proposed hypothesis invites testing. This testing involves variables identified as follows:

Controlled variables - Variables that are held constant through the experiment so as not to influence the results.

Manipulated variable - The variable that is purposefully changed in an experiment.

Responding variable (dependent) - The variable that changes in response to the manipulated variable.

A Trial Run for Variable Identification

🔺 A student desired to test which of two colors of light would produce the best lima bean plant growth. The student selected three identical seeds of the same mass and size. These were planted in identical soils, containers, depth of soil, and provided with the equal amounts of sunlight, water, and temperature. Plant A was covered with a red cellophane material. Plant B was covered with a blue cellophane material. For an experimental control, Plant C was covered with a clear cellophane. The growth of the plants was recorded daily. The student graphed the results and interpreted the data. A conclusion was made supporting or refuting the original hypothesis.

Make an inference relative to the original hypothesis for this action.
State the controlled variables.
State the manipulated (independent) variable.
State the responding (dependent) variable.
What experimental control did the student utilize?

***Hypothesis:

Lima beans grown under colored cellophane will grow faster than under clear cellophane.

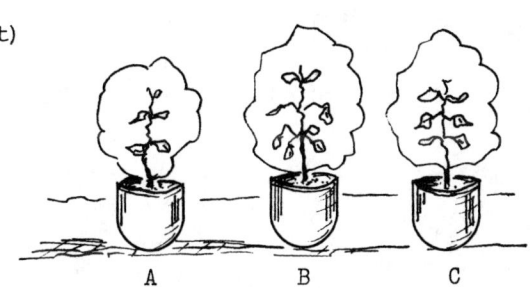

Is this the only inferred hypothesis that can be made relative to the described action? Create another suitable hypothesis.

Controlled variables are:

Seed size and mass	Water	Soil
Containers	Depth of soil	Sunlight
		Temperature

*
The manipulative variable was the colored cellophane.
The responding variable was the growth of the lima bean seed.
The controlled experiment was Plant C using clear (white) cellophane.

▲ Given the following: an iron nail or bolt about 5 cm long, insulated copper bell wire (1 meter), paper clips, one 6 volt dry cell battery, wire strippers, a compass, and a stated hypothesis, "The strength of an electromagnet increases if the number of turns in the coil are increased." State the following:

The controlled variables, the manipulated variable, and the responding variable.

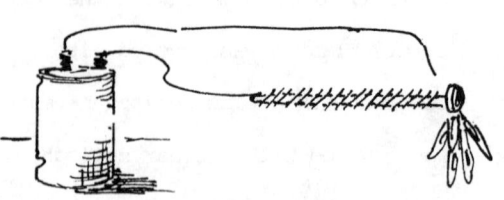

* An electric current flowing through a wire produces a magnetic field. This attribute can be used to construct an electromagnet. The strength of the electromagnet can be increased by more turns about the iron core or by adding additional dry cells.

▲ Given the following: honey, spaghetti, beets, jelly, noodles, bread, potato, soup crackers, tomato, corn flakes, orange, sugar, and a stated hypothesis, "Carbohydrate foods with starches in them react to an iodine solution." State the following:

The controlled variables, the manipulated variable, and the responding variable.

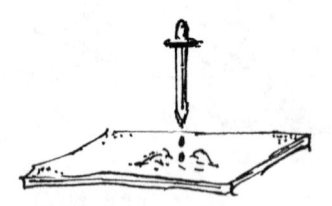

* Starch is manufactured by plants and animals as a means of storing sugar. A simple test for starch can be made by placing a few drops of an iodine solution (2-3 drops per 25 ml of water) on food. If starch is present, the iodine changes from a reddish-brown to a blue-black color. Some view this color as a purplish blue.

▲ Given the following: two potatoes, knife, plasticene clay, a balance, weights, and a stated hypothesis, "A peeled potato will lose more of its water than an unpeeled potato." State the following:

The controlled variables, the manipulated variable, and the responding variable.

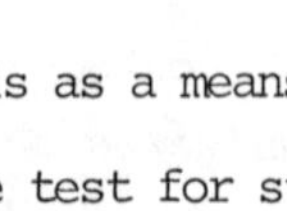

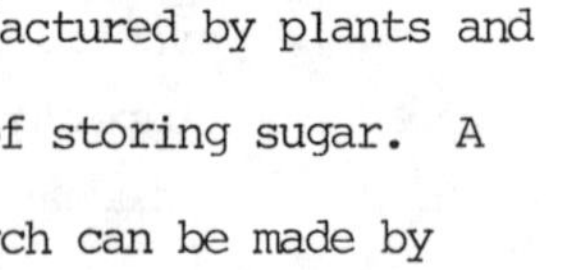

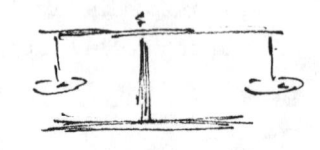

* Evaporation from root systems is slowed down by the presence of an outer skin or protective layer. While roots serve primarily for anchorage and absorption of soil water, the root epidermis permits water to enter the root by osmosis. This same skin or outer covering acts as a barrier for the escapement of interior root water.

🔺 Given the following: ice, water, plastic container, a balance, weights, and a stated hypothesis, "When ice melts, it contracts and its volume decreases and its mass remains unchanged." State the following:

The controlled variables, the manipulated variable, and the responding variable.

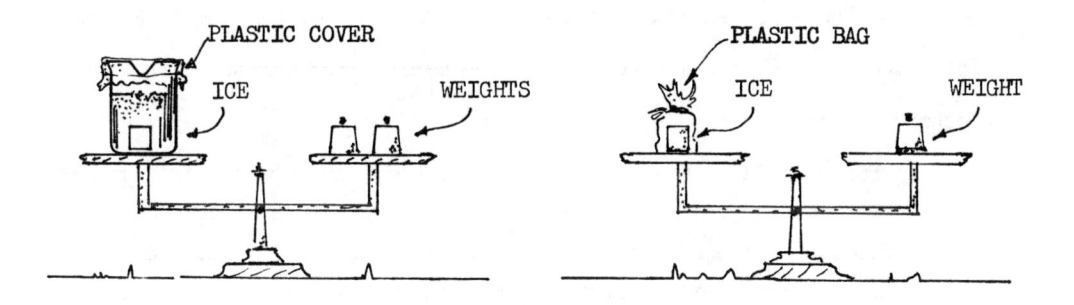

Try it! Can the stated hypothesis be supported?

Try it! Can the stated hypothesis be supported?

🔺 Given the following: salt, water, plastic vial, a balance, weights, and a stated hypothesis, "Salt dissolved in water contained in a vial weighs the same as the sum total of each item prior to the dissolution." State the following:

The controlled variables, the manipulated variable, and the responding variable.

🔺 Given the following: test tube, clear plastic from a garment bag, a colloidal dispersion of starch and a glucose solution, a container, water, iodine, and a stated hypothesis, "A permeable membrane can be selective in allowing materials to pass through it." State the following:

The controlled variables, the manipulated variable, and the responding variable.

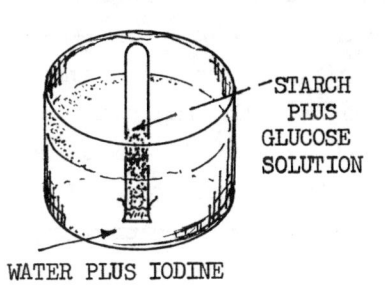

* Positive results will appear in about 20 minutes. The contents of the test tube will turn blue while the solution in the container will appear not to have changed. The blue coloration in the test tube is due to the iodine in the water of the large container having been diffused through the "membrane" into the test tube, but the starch is unable to diffuse into the water of the large container. If the solution in the large container is tested for sugar, it will be found that some glucose has diffused into the beaker from the contents of the test tube.

△ Given the following: straws, limewater (commercially available at drugstores), a timer, quart glass jar, permanent marker, and a stated hypothesis, "Exercise increases the production of carbon dioxide." State the following: The controlled variables, the manipulated variable, and the responding variable.

* Carbon dioxide, a waste product of the body, is expelled in relation to its generation. A greater intake of oxygen results in a greater output of carbon dioxide. Limewater is an indicator for carbon dioxide. Comparisons can be made at rest, mild exercise, and vigorous exercise. In all three cases one's exhalations can be funneled through the straws and bubbled through limewater. Exhaled carbon dioxide bubbled through limewater will make the limewater turn milky as calcium carbonate is formed. The passage of excess carbon dioxide through the limewater will make the milky appearance disappear.

△ Given the following: one 6 volt battery, wire, steel nail, paper clips, and a stated hypothesis, "Increasing the diameter of the steel core (nail), keeping the number of wire wraps the same, diminishes the strength of an electromagnet." State the following: The controlled variables, the manipulated variable, and the responding variable.

* Use the alligator clips attached to both ends of the insulated copper wire. A forty cm length of wire works well. Keep the copper wire wrapped snugly about the steel nail. The wire will feel warm, but there is no danger. Vary the diameter of the steel nail. Keep the length of the nail constant. At the conclusion of each core variation, test the strength of the electromagnet by collecting and counting the number of paper clips the electromagnet picks up. Record the data. Plot the data on a graph. Interpret the data to support or refute the hypothesis.

ACTIVITIES TO ENHANCE THE PROCESS SKILLS OF HYPOTHESIZING AND EXPERIMENTING

Hypotheses are generalized statements rising from confirmed, frequent observations about relationships between variables. A hypothesis is a statement which when responded to, directs the experiment. Sometimes a hypothesis has been described as an "educated guess." Some definitions include descriptions such as "an if...then statement" or "a statement to be disproved." In reality a hypothesis is a prediction in which cause (independent variable) and effect (dependent variable) are clearly stated. For example,

If the mass of the metal washer is doubled then the bouyance of the plastic container into which it is placed in water is reduced.

This statement communicates exactly what will be done and the expected effect of such action. It leads the experimenter through the experiment.

⚠ The Situation:

Given: White bread, blue cheese, water, clear plastic bags with twist ties, magic markers, magnifying glass, sunlight, thermometer, refrigerator, and brown paper bag.

Study the list of materials. What might you infer would be a suitable hypothesis that could be tested involving this material? What would be the controlled variables? The manipulated variable? The responding variable? If you wish, conduct an experiment based on your hypothesis.

*   Hypothesis: If the temperature is increased in the presence of moisture, then the rate of decomposition is increased.

The variables held constant would be the amounts of bread and cheese, plus water placed in three separately marked (A, B, and C) plastic bags closed with twist ties, and each provided the same amount of sunlight. The manipulated variable is the amount of heat supplied to each sample. The responding variable would be the growth of the mold.

*** Bacteria, molds, and yeast are living microorganisms. Microorganisms cause decay and spoilage. They are called decomposers. They decompose or break down, complex dead materials into simpler materials. Moisture and increased temperature enhances decomposition

Caution: Once the process of decomposition starts, do not open the sealed plastic bags, don't taste, touch or smell spoiled food. When done with the observations, comparisons, and conclusions, throw all unopened bags away. Most, perhaps all, of the microorganisms that will grow are harmless but some could cause allergic reactions, an infection, or even serious illness.

⚠ The Situation:

A one square decimeter piece of stiff cardboard was wrapped in aluminum foil. One surface of the square was coated with petroleum jelly. This was placed outdoors in an undisturbed area, open to the sky for a twenty-four hour period. This square was later examined for material trapped in the jelly. What might you infer the hypothesis would be that would generate such action? What would be the controlled variables? The manipulated variable. The responding variable? If you wish, conduct an experiment based on your hypothesis.

*   Hypothesis: Micrometeorites or extraterrestrial debris strikes the earth and can be collected and measured. The amount collected varies with the geographic location.

The variables held constant would be the collection surface area, the amount of jelly on the exposed surface area, and the time. Weather would be a variable that would be difficult to control. If possible, similar conditions should be adhered to. The manipulated variable would be the geographic site selection for the placement of the aluminum squares. The responding variable would be the various amounts of micrometeorite material recovered at each site.

*** The micrometeorites collected will be of Ni-Fe composition whose origins were most likely in the asteroid belt, as differentiated from "shower" meteors whose composition is considered to be mostly CO_2 and hydrogen compounds. While there will be much uniformity as to the number of micrometeorites, geographical variations in the distribution of the collection squares plus wind speed, protection from trees, and the like will make for some differences.

The surface of the earth has been calculated at 5.10×10^{18} cm^2 or 5,100,000,000,000,000,000 square centimeters. A square decimeter is equal to 100 square cm. Dividing the area of the collection square into the total area of the earth would provide the earth's surface area in square decimeters. Multiplying this number times the count of the number of squares will provide a total earth collection over a 24 hour period.

⚠ The Situation: One Potato, Two Potatoes, Three...

Given: A dozen potatoes, 24 metal paper clips, a five volt light bulb in a holder, copper wire cut into 13 lengths (bare copper wire is preferred otherwise, the insulation will need to be stripped back about one-half an inch at both ends of the 13 lengths), twelve strips each of zinc and brass cut into strips about 2 inches by 3/8ths of an inch. Brass strips can usually be found in hobby shops that cater to boat and airplane model builders. Zinc strips can be found in hardware stores, lumber company warehouses, or scientific supply houses.

Organize the potato, electricity generating station as follows:

With the same general orientation, insert one brass strip and one zinc strip facing each other into each potato placed about 3/8ths of an inch apart. Connect the wire to the paper clips by winding several turns of wire about each strip. Attach the paper clips to the strips. Connect the wire from the brass strip in one potato to the zinc strip in another potato in chain-like fashion completing the circuit. Fully connected, the light should go on.

This is an activity. Turn it into an experiment. State a hypothesis. Identify the manipulated variable. The responding variable. Carry out your experiment. Collect the data. Graph the data. Interpret the data. Conclude whether you support your original hypothesis or refute it.

Sample hypotheses:
  A. Increasing the distance between the brass and the zinc strips will result in increased electrical energy.
  B. Decreasing the number of potatoes in the system will result in a decrease in electrical energy.

The Situation: Bouyancy

Materials: 35 mm plastic, film containers (clear containers are best for marking graduations on their exteriors), clear plastic trays, water, metal washers, and a balance.

Weigh the metal washers. Accumulate at least six washers of equal weights (or nearly equal). Use three, 35 mm containers and mark or tape a strip of graph paper on the outside of each. The graph paper should be graduated and marked in consistent units. Cover the strips with clear tape and mark the containers A, B, and C.

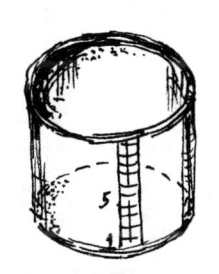

What is the volume of a 35 mm film container?

Empty, plastic film containers will not float upright in water. Place one washer in container A. Place two washers in container B. And, in container C place three washers. It is not necessary to replace the caps on the containers. Make an inference relative to the bouyancy of each container when they are placed upright in the water. Which container would you infer will float highest in the water? Lowest? Validate your inference by placing all containers in the water. Record your observations. Collect observable data. Construct a graph. Interpret the data. What do you conclude? Inasmuch as no original hypothesis had been stated, your conclusions are after the fact. Surely you will have discovered, uncovered, or searched out some knowledge, but its relevance may be in question. It is more efficient and productive to establish your hypothesis first and search for the evidence to support or refute your initial hypothesis.

A Sample Hypothesis:
If the density of the liquid in which the containers are placed is changed, the bouyancy of the containers will be changed.

Other hypotheses can be formulated relative to maintaining the same density and altering the container weights further, altering the volume of the containers by removing or shortening the sidewall of the containers, or by clustering more than one container.

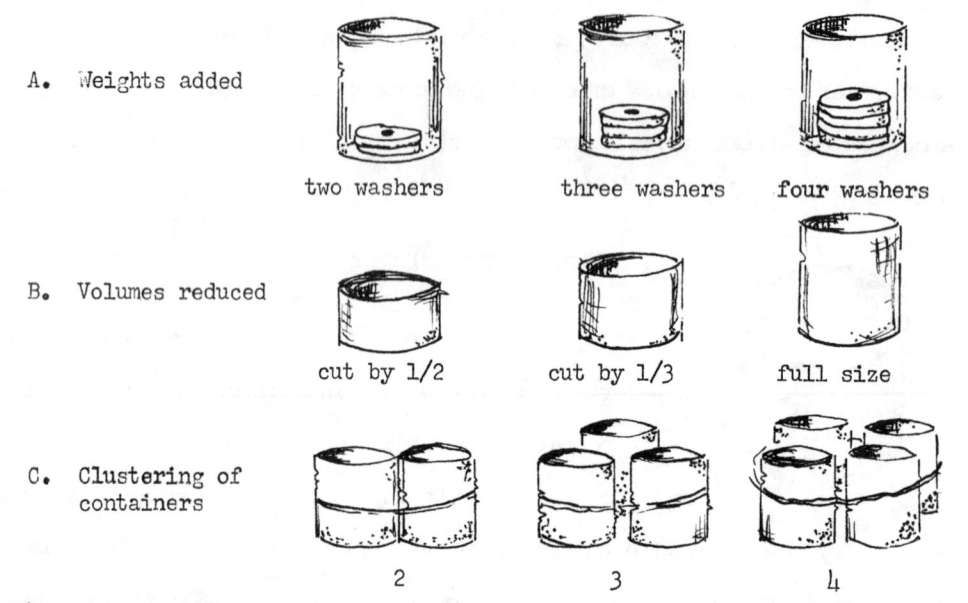

A. Weights added

    two washers   three washers   four washers

B. Volumes reduced

    cut by 1/2   cut by 1/3   full size

C. Clustering of containers

    2   3   4

⚠ In place of metal washers, fill containers A, B, and C with unequal amounts of water. How does this effect bouyancy? Formulate a hypothesis and conduct your experiment to support or refute your hypothesis.

⚠ Completely fill one container with water. Place this in a container which has sufficient water to cover the 35 mm container if it sinks to the bottom. Observe. If your container is not completely filled, add water to it while it is floating in the large container. Why does the 35 mm container not sink?

*   An object will float, irrespective of its density, if it is lighter than the weight of an equal volume of water. The plastic material of the 35 mm container is lighter than the amount of the volume of water that this material displaces.

⚠ The Situation: Crushing and twisting (or compression and torsion)

How many books will each 3 x 5 card structure support?

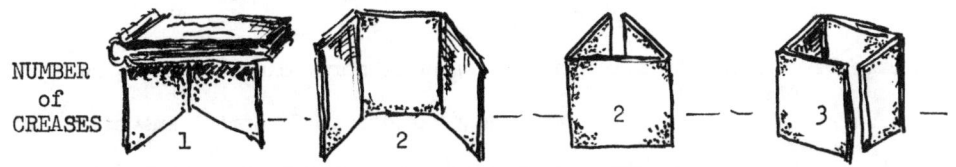

NUMBER of CREASES: 1 — 2 — 2 — 3 —

How might each structure be altered to enable it to support a greater weight?

Does overlapping a joining crease alter the support ability of the structure?

Does changing the geometric shape alter the structural strength of the structure?

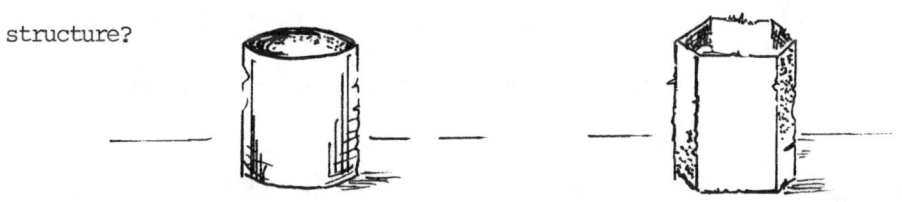

Does the addition of internal crossmembers alter the structural strength of the structure?

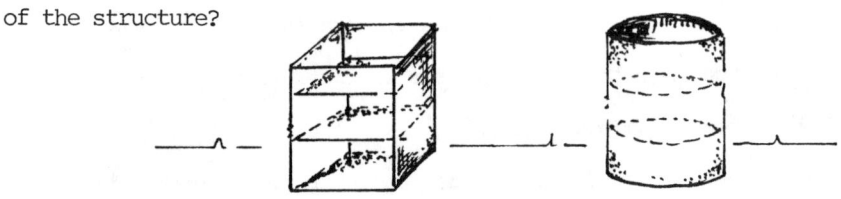

Have you conducted an experiment? Your response depends on your definition of an experiment. Review your actions. Did you state a hypothesis? Did you identify your manipulated and responding variables? Did you collect, plot, and interpret accumulated data? Was an experimental control necessary? What did you conclude?

OPPORTUNITIES FOR INVESTIGATION

<u>Biology</u>   Premise: Can the juice from crushed bugs mixed with water control bugs in your garden?

Some Direction: An all-purpose bug juice spray can be readily made by collecting as many varieties of harmful insects in your garden as you can find. Using a plastic container and an electric drill with a paint mixer attachment inserted into the bit, grind the bugs up (an old blender would work easily as well). A mortar and pestle would work also. To each one-half cup of bugs, add two cups of water and liquify. Strain this through a piece of cheesecloth so that your sprayer will not become clogged with bug particles. To each cup of this mixture, add four to eight parts of water. The spray remains effective for a long period of time...in some cases up to two or three months.

<u>Horticulture</u>   Premise: Plants planted in root-restricting pots experience lower yields.

Small pots tend to stunt growing plants, much as drought does. Unlike drought-related stunting, the stunting from root restricting small pots is not caused by decreased photosynthesis.

Some direction: Given adequate water and nutrition, root-restricted plants conduct comparable photosynthesis and even offer higher yields per volume of soil than non-stressed plants in large pots. Small pots are almost twice as efficient at producing tomatoes. The <u>key</u> is adequate water and nutrition.

Horticulture   Premise: The color of the light a mulch reflects back onto a growing plant can significantly affect growth, shape, size, and productivity.

Gardeners are encouraged to mulch their plants to reduce weed growth and moisture loss in the surrounding soil. A variety of mulch materials are used, for example, shredded newspaper, bark chips, straw, grass clippings, etc. Some suggest opaque plastic material for its ability to collect and retain heat.

Some direction: Red mulch when used with tomatoes gave larger fruit and even increased the total number of fruit.
It is suggested that some plants may even have their own preferred color.

Physics   Premise: A burning candle flame will have its profile (tear shaped) altered in the absence of gravity.

Some direction: A flame from a candle is a relatively simple, efficient way of transferring large amounts of energy to a specific location. Convection currents bring fresh chemical fuel into the flame's luminous combustion region and carry away hot products giving the flame its distinctive tear-drop shape.

In the absence of gravity, no convection occurs and flames are spherical. Eventually the flame suffocates under a blanket of their own products. How might a flame in space be kept burning?

Electroculture   Premise: Plants will grow better and be more productive
when grown within an electrically charged atmosphere.

Some direction: Metal is an excellent conductor of electricity. Metal, in the form of wire, can be used to attract electricity to vegetable gardens. String copper wire from metal posts set at ends of rows of plants. The wire should stretch closely above the plants but never directly touching them. The wire should be moved up as needed to accomodate growth in the plants. The wire in its new position should not touch the plants.

In many cases the yields have been increased by as much as fifty percent.

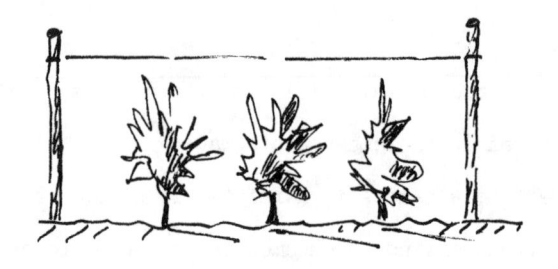

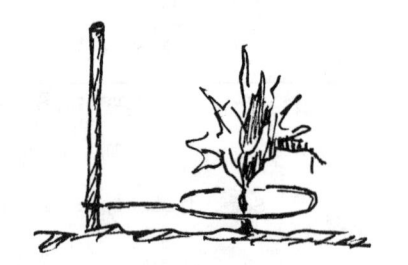

And, in addition: Electrically charged atmospheres can be constructed around individual plants. Drive a metal rod into the ground beside the plants. The wire should be fastened to the stake. The wire should form a circle about the stem of the plant, but it should not touch the plant. Compare the individual plant's productivity and growth to individual plants when the linear, copper wire distribution method is used. Which is the preferred method?

<u>Physics</u>   Premise: Adhered common adhesive tape when peeled, in the dark, from a pane of glass emits a faint blue or blue-white light along the line where it is separating from the surface.

Some Direction: To observe the soft glow of light, accustom your eye to the dark for a period of at least ten minutes. Then position your eye close to the separation area as the tape is pulled away from the glass surface.

The area under the tape is a cohesive/adhesive zone wherein electrons move across the interface of the two materials (tape and glass). This is not usually a uniform development. Areas of unequally charged accumulations of electrons exist, either positively or negatively charged. When the tape is peeled away, these unequally charged zones are altered. When critical values are reached sparks jump through the air exchanging charged particles to establish some level of equality or neutrality.

As the tape is peeled away air rushes in. Particles collide with the air's gas molecules, leaving the molecules in an excited energy state. In the process of the molecules returning to the unexcited state they emit some light that is visible. Additional light may be emitted by the surface of the glass or the tape when they are struck by the particle streams.

Horticulture   Premise: Red kidney beans, when planted, are uneffected in terms of productivity in responses to the direction their tendrils coil about a vertical support pole.

Some direction: Red kidney beans are planted and their tendrils guided in a clockwise manner about a support pole. This is compared to similar, red kidney beans planted and their tendrils guided in a counterclockwise direction. Compare growth and yield. Some interesting results can be obtained. Try it!

Physics   Premise: The center of gravity of an irregularly shaped object can be readily located.

Some direction: Finding the balancing point of an irregular shaped object can be accomplished by hanging a plumb bob or weight suspended from a string from different points of the object. The point of intersection of these lines is the center of gravity for this particular shaped object. A pin stuck through the shape at this point would permit the shape to rotate and stay in whatever position it was turned to. Find the center of mass of a metal coathanger.

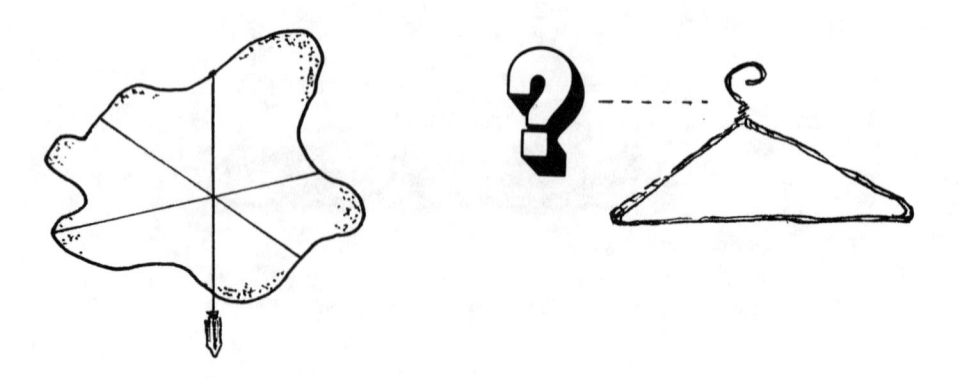

Biology   Premise: Will a pollutant (detergent water) effect seed germination?

Some direction: This involvement is an attempt to show that the addition of increased concentrations of a pollutant diminishes seed germination.

Plant seeds, either mung, lima, or radish seeds in five different containers. Label one container as the "control," and the remainder as A, B, C, and D. Keep everything constant for all containers. This should include the type of soil, the quantity of soil, the depth of planting, seeds of equal variety, shape, mass, initial watering procedures to sustain growth, temperature, and exposure to light. Establish the amount of liquid to be given to all plants and the watering cycle, for example, 30 mL of liquid every other day. This amount should be decided by you to fit your particular situation. The control plant should receive the same amount of liquid as provided the other plants minus the pollutant and dispensed in the same time cycle. The manipulated variable is the concentration of the pollutant. For example, plant A should receive one drop of detergent per 30 mL (approximately two tablespoons) of water. Plant B should receive two drops of detergent per 30 mL of water, and so forth. Formulate a hypothesis and carry out your experiment. Observe the reactions of the various plants to the various treatments. Collect the observed data. Record the data. Interpret the data. Conclude whether or not the data supports or refutes the original hypothesis.

Geology   Premise: Can radioactivity be used as an age determinate
                          (a means to reconstruct past geologic time frames)?
Some directions: Radioactive elements disintegrate at a uniform rate un-
                  effected by the surrounding conditions of temperature,
                  pressure, or geologic disturbances. Radioactive ele-
                  ments become an ever-accurate perpetual clock. A con-
                  crete representation of this phenomenon can be made.
                  Start with an eight inch square sheet of paper. All of
                  the area of the square is presumed to be radioactive.
                  Fold the paper in half. Tear the paper along the fold.
                  You now have two halves. Consider one of the two halves
                  to be radioactive and the other half transformed into
                  another element, for example, lead which is not radio-
                  active. Assign a time frame to this event, for example,
                  1,000,000 years for this separation to occur. Fold the
                  radioactive portion of your paper (this is the half por-
                  tion of the original eight inch square sheet). Tear the
                  radioactive portion along the fold forming two halves
                  (each of which is one quarter of the original eight inch
                  sheet). One of these halves is considered to be radio-
                  active. The other half has presumably disintegrated in-
                  to non-radioactive lead. The time for this separation
                  or division to have occured would also be 1,000,000
                  years. Continue this process, always folding and tear-
                  ing the radioactive half into two sub-halves. The time
                  span for each disintegration into lead would be

1,000,000 years. This span of uniform time for the changing of any of the entire radioactive portions to where they are reduced to 50 percent lead is called the half life of the radioactive element which is disintegrating. Knowing this uniform rate or half life, and knowing the current status of disintegration, one can reconstruct or work back by adding the number of half life years, and thus establish a time frame. For example, if one started with an eight inch square sheet of paper and you ascertained that you currently had a piece of paper that had the dimensions of 4 inches x 2 inches, the necessary folds and tears necessary to arrive at that point from a full 8 inch square could be calculated. A 4 inch x 2 inch rectangle would have taken three folds and tears or related to time, three, 1,000,000 year time spans for a total period of 3,000,000. The half-life for Carbon 14 is 5,730 years. Construct a graph plotting time in half-life increments versus the percentage of the original amount of the radioactive material. Do this for either Carbon 14 or Uranium 238. The half-life for Uranium 238 is 4.5 billion years.

Biology   Premise: Will increased sunlight effect the amount of plant transpiration?

Water exists in a continuous column from the root tips of a plant or tree up to the leaves. Water is lost through the stomates in the leaves. This process is called transpiration. As this water is lost through transpiration in the leaves, a force acts on the column of water in the plant. This pulling force causes water to move up the xylem from the roots.

Some direction: Three plants with good leaf foilage are needed. Geraniums or well grown, lima bean plants will do nicely. Label the potted plants. Wrap each plant and pot entirely in clear cellophane. Place one plant in direct sunlight; one in indirect sunlight; and one in darkness. Observe what happens. Is this a good experiment to test the above premise? Not really! The variables have not been fully controlled. We have manipulated the sunlight, but the sunlight in turn manipulated the temperature within each plant's cellophane envelope. Hence, we do not know what to attribute any observed results to. Were the observed changes due to sunlight amounts or changes in temperature from the sunlight? This is a good example of confounding the variables. If sunlight is to be the manipulated variable, then temperature must be controlled. If not, the premise should be altered. Formulate a hypothesis and carry out your experiment.

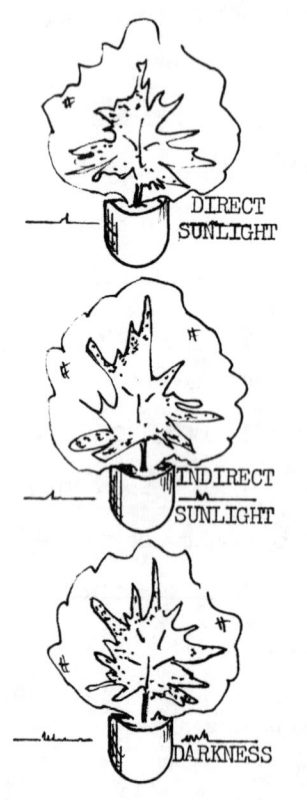

This involvement could be expanded considering transpiration as influenced by the area of a leaf, what type of leaf transpires most, and does the age or location of a leaf on a plant make a difference in the transpiration process.

Physics   Premise: How does an applied force effect the movement of
               objects of increased weight?

Some direction: Construct a trough large enough to accomodate the object that you select to roll down the inclined plane. A one-inch diameter steel ball (about 70 grams) will do. Golf balls, large marbles, or plastic spheres that have sufficient mass can also be utilized. In this instance, the mass of the rolling object remains constant as does the angle of the inclined plane. The manipulated variable is the weighted object placed at the bottom of the inclined plane. This object's weight is sequentially increased with subsequent rolls down the inclined plane.

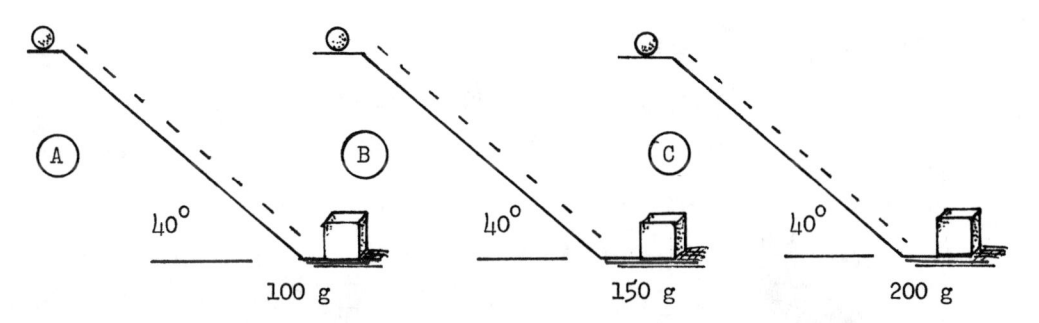

A	B	C
40°	40°	40°
100 g	150 g	200 g

Additional hypotheses can be written by considering other variables. If the mass of the object at the base of the inclined plane is kept constant, the weight of the rolling sphere can be manipulated. Or, the weight of the rolling sphere and the mass of the object at the base of the inclined plane can be kept constant and the angle of the inclined plane can become the manipulated variable. Formulate a hypothesis of your choice and perform the experiment.

Botany   Premise: Seeds can be germinated at any time of the year.

Some direction: Seeds need heat plus water to grow into plants. Light enhances robust growth. If this is true, why do seeds in apples, tomatoes, watermelons, etc., not grow into plants while still inside the fruit itself. Heat and moisture are present. Seed growth is tied to seasonal cycles. Inhibitors within the seed block growth until appropriate seasonal cycles have been passed through. Plants can be fooled into by-passing seasons, for example, the placement of avocado seeds in a refrigerator for a period of time to mimic the passing of the winter season. Try by-passing a season by falsely leading a plant seed to assume that it is now the season for growth.

Chemistry   Premise: A liquid can be systematically diluted (serial dilution).

Some direction: Prepare a glass of water that contains just one eighth of a drop of ink in it. This, at first glance, would appear to be a difficult task. This can be accomplished by serial dilution. Use a dropper to place one drop of ink into a glass of water. Stir the ink into the water vigorously. Pour half of this mixture out. Add more water until the glass is filled again. At the moment you have a glass of water with one-half a drop of ink in it. Repeat this process twice more. This will take you to one-fourth of a drop per glass and then the final dilution of one-eighth of a drop per glass of water. How would you proceed to prepare a glass of water that contained one-tenth of a drop of ink in it?

Biology   Premise: Can color aid or hinder recognition of living things (mimicry)?

Some direction: Within a five-meter square plot of grass-covered ground, scatter fifty, green-colored toothpicks (natural toothpicks soaked in green, food coloring for a 24 hour period and then allowed to dry work nicely). At different times of the year and at different geographic locations, different colored toothpicks compatible with the conditions are recommended. Within the same measured plot of ground, mixed in with the previously scattered green toothpicks, scatter fifty, natural-colored toothpicks. Later, have a non-observer to this act, collect as many of both variety of toothpicks as possible. Graph the results. From this activity formulate a hypothesis. Carry out your experiment.

Biology   Premise: What happens when a population exceeds the supply of an essential ingredient needed for growth?

Some direction: All living things compete for the essentials for life. Each living thing has an optimum requirement for food, air, water, space, etc. Any deficiencies in these requirements cause the living organisms to suffer in one way or another. If availability levels of these essentials for growth drop too low or are not available, living things may die. Plant lima beans in several containers. Vary the number of seeds from one to one hundred in different containers. Formulate a hypothesis. Identify the manipulated variable. The responding variable. Collect the observable data. Does the data support or refute your original hypothesis?

Health Premise: Physical body action deteriorates with the duration of the time involved.

Some direction: The body responds to the environment. As increased demands are placed on the human body, it reacts accordingly. Prolonged action, with insufficient recuperative time, has a deleterious effect on an individual's performance. How many jumping in space, deep knee bends, situps, pushups, etc. can you do? How does the number of actions per unit of time change over time? Quantify the results. Construct a graph of the data to organize the data. Interpret the results. What happens to performances sustained over periods of time?

Botany Premise: Measuring leaf transpiration.

Some direction: Water exists in a continuous column from the root tips of a plant or tree up to the leaves. Water is lost through the stomates in the leaves (transpiration).

Is leaf transpiration increased by elevated temperatures?

Put some water in a glass. Determine the weight of the glass and water. Cut a small hole in a cardboard square large enough to allow a leaf stem to slip through and into the water. Enclosing a thermometer, place another glass over the leaf to form a closed container. Make two, similar setups determining the weight of each. Label these A, B, and C. Place these in areas where the only variable is temperature. At intervals record the temperatures. At some point in time, separate the top portion from the lower portion. Reweigh the glasses and the remaining water. Does water loss through transpiration correlate with variations in temperatures?

APPROACH 3:   **THE IDEATION - GENERATION APPROACH** - Generation by Expansion through Creative Action

This approach can augment the morphological approach and/or the process approach. Or, it can be used in isolation. The ideation-generation approach is associated with the creative act of brainstorming. Brainstorming is the act of stretching the mind to generate new ideas to assist in the solution of problems. Ideation-generation is a way of looking at things in a somewhat bizarre fashion to generate ideas that fall beyond the realm of common-place thoughts. It is hoped that among these sometimes outlandish, contrary, or bizarre ideas may lie the germ of fruitful unique thoughts worthy of further investigation.

Some simple rules for exercising the act of brainstorming are:

- Entertain all idea contributions. No considerations are to be judged "right" or "wrong," "good" or "bad." Do not criticize any idea. All ideas are acceptable.

- Collect as many ideas as possible. The momentum of the productive generative moment should be allowed to run on unabated. Quantity, at this point, is paramount to quality. Quality judgements can come later.

- Think in a cavalier manner. Be audacious! Often times, the most fruitful idea come to the surface when bravado is substituted for caution. High flying ideas can be re-engineered later.

- Zipper your ideas together. Unite two or more ideas (yours or someone else's) into one new one.

- Productivity is infectious. We feed off one another. One person's idea sparks an idea in another person's mind. Piggy back on other people's ideas.

Creative Action for Ideation - Generation

    Expanding Ideas:

        OPTIMIZE   Make the most of what you have. Recognize your ability, your potential for greatness. Observe with purpose. Develop not a simple curiosity, but a rage to know. Think positively. Capitalize on your subconscious mind. Ask provocative questions. Relax. Associate Ideas.

Getting the most out of what you have means exercising your ability to think. Thinking is the forgotten process skill. It is the most productive activity a human being can undertake. Our society is fueled by ideas. Sometimes we get too busy to think. Thinking takes time...relaxed time. A person in deep thought is not always an elegant sight. And, because of this thinking is sometimes looked on as suspiciously like loafing.

        ADAPT   Grab at ideas. Stretch for them. Search for solutions to problems. Ask yourself "What else is like this? What other ideas does this suggest?" Learn from others. All of us is better than any one of us. Emulate your mentors. Then, cut yourself away and, on your own, run with the wind. Look for the "Eureka" moments. Invent something.

            Example: Inventions usually arise from an irritating situation and a search for a preferred way to accomplish something. For example, photographic film - once one inserts a specific type of film with a fixed-rated ASA speed into a camera, one is stuck for the duration of that roll with the same film speed. Is it possible to have a base film that could by some manipulation of the camera, have its film speed changed to make it more flexible for specific camera shots? Or, 30 credit cards- why not one and, some speedy way to verify its credibility at the time of use. Or, dripless paint cans to avoid a cleanup. Inventions can arise from a need for a product, for example, a dry packaged, non-flammable gas (just add water and it becomes useable), or, ...

MODIFY — Take old, pre-existing things and give them a new twist or tilt. Change the purpose. Change the packaging. The taste. The shape.

  Example: What does one do with old TV cabinets? Some people throw them out. Some make sewing cabinets, liquor cabinets, etc. out of them. What can you do with the old TV tubes?

MINIFY — Make it smaller. Omit something. Streamline it. Split it into several subcomponents.

  Example: The food industry has capitalized on minification, for example, salt free, sugar free, light beer, portions for two, portions for one, etc.

MAGNIFY — More! Take something and make it larger. Taller. Heavier. Double the dimensions.

  Example: Industry is constantly using this concept. Larger containers of soap, bigger pizza, bigger hamburgers, three scoops instead of two, higher, faster amusement rides, etc.

SUBSTITUTE — What else can be used in its place? Use substitute ingredients or materials. What can do the task for less cost?

  Example: Generic brands instead of name brands of consumer products. Fiber glass car bodies instead of metal bodies. Plastic instead of wood. Man made fibers instead of duck down, etc.

REVERSE Transpose positive and negative aspects. Try opposites. Turn it backwards. Look at things as though the world was upside down. Alter your perspective.

Example: Consider these reversals:

What would the world be like if the sun rose once every two days instead of every day? Once a week? What would the world be like if the land masses were occupied by water and the water areas occupied by land masses? How would this change things? What would the U. S. be like if its mountain ranges had an east-west trend instead of a north-south trend? What would the U. S. look like in relation to growth and settlement patterns if the Pilgrims had landed on the west coast at Los Angeles rather than on the east coast at Plymouth Rock?

REARRANGE Alter components for a new arrangement or design. How can we make it better by making it differently? Use new designs.

Example: Bicycles have changed little since their early conception. Recent competitive contests have spurred a host of new fangled contraptions, some of which are pedaled from almost an inclined position to lying prone and using one's arms to propel the vehicle. Different drive trains are utilized to gain more speed. Parts are eliminated. Some bicycle designers have added wind deflectors to reduce air friction, etc.

COMBINE Blend ideas. Combine ideas. Combine uses. Add something to the existing. Take the best of many things and recombine them to facilitate a new end result.

Example: If one could combine plant parts such as a carrot bottom with a cabbage head, it would be a nice accomplishment to depict combining. Redesign a new animal utilizing the best attributes for survival garnered from any and all animals of the earth. What would it look like? Design a new musical instrument incorporating the best of several. What would it look like? What would it sound like?

NEW USES   New use for the item as it is. New uses stemming from a modification of the old.

Example: Old empty paper milk containers make fine flower pots. Discarded jars or cans make suitable receptacles for many items. Old tires have been cut into soles for shoes. Large plastic leaf bags, by cutting and slitting them, can readily be transformed into rainproof, plastic poncho-like garments.

AN EXAMPLE: BRAINSTORMING with a frizbee

To reinforce the notion of idea expansion through creative action, one can OPTIMIZE one's ability to generate ideas by applying one's thinking to almost any common object. Using the Ideation - Generation Approach new ideas can be generated, for example, by considering a common object such as the plastic, recreational toy, throwing disc called the "frizbee" as follows:

ADAPT   Frizbees are slightly convexed plastic dish-like objects. Inverted, they can serve as platters, snow scoops, or paddles. They can be used as a super-sized, jello mold. By the addition of pliable, plastic handles located in their centers, one could use frizbees as body shields or as swimming fins. They can be fastened on to one's feet, much like snowshoes, to assist in walking in deep snow. Frizbees can even be used to pan for gold.

MODIFY With holes punched in them, frizbees can serve as sieves or strainers. Frizbees can be used as a bird-feeder cover or to discourage squirrels from eating from a bird-feeder. The frizbee can be used as a mold for making small, circular, concrete, garden-stepping stones. Also frizbees can be used as a pizza-serving tray or a snack dish holder. Pierced by a spindle, it can serve as a spinning top. If stiff enough, frizbees can be used as wheels for small toys, or carts.

MINIFY Frizbees can be made smaller. Their size can range from small, plate size to silver dollar size. These would be mini-frizbees. The convexness of the disc can be reduced. The size of the curved lip can be made smaller. The mass of the frizbee could be reduced while maintaining the original dimensions of the frizbee.

MAGNIFY One could make a monster size frizbee ranging from normal size to garbage-can cover size. Maintaining the original dimensions, the mass could be increased. Increase the convexness of the disc. Give the curved lip greater depth.

SUBSTITUTE Use old phonograph records, aluminum pie plates, or when desperate, an aged, dried, buffalo chip in place of the frizbee. Try a cardboard frizbee. Papier Mache. Try an old, metal coat hanger fashioned into a circle and covered with kite paper. Construct a balsa wood circular frame with the necessary curved balsa cross sections. Cover this with stretched cellophane or cling-wrap material fashioned into the classic frizbee shape.

REVERSE Glue two frizbees together in a double convex, clam-like manner. Glue two frizbees together in the reverse manner forming a double concave shape.

REARRANGE Change the shape from a circular disc to a triangular (or any other geometric shape) shaped object. Redesign the disc so that when the object is in flight air passes through small holes cut in the cover. Air so directed over a musical toy (similar to a harmonica) can generate a tune. Frizbees may be further rearranged by drilling small pop-up sections in the cover section which house small parachutes. These pop-ups and their parachutes are activated when the disc is in flight. Upon reaching maximum height and with the frizbee's descent initiated, the pop-up portholes are activated and the disc is parachuted over a preselected site. Also, the frizbee objects could be painted with luminescence paint for night-time play.

COMBINE Combine frizbee throwing with other sports, for example, frizbee baseball, frizbee football, frizbee golf, or frizbee basketball. Use flying frizbees for moving targets for archery, skeet shooting, or targets for other flying frizbees.

NEW USES Redesign the disc so that it can be retrievable. Then, the frizbee can be tossed into inexcessible areas or over the water to accomplish a specific purpose such as broadcasting seeds, etc., then retracted by a fine filament line. Frizbee design applied to new aircraft design, utilizing vertical take-offs and landings.

A new boomerang frizbee that always returned to the thrower.

To augment the notion of idea generation as put forth by the three generative approaches, consideration should be given to those common variables which permeate science. These variables are broadly applicable, in every instance. Some of these variables are:

    light and color

    temperature

    pressure

    mass

    volume

    surface area

    density

    humidity

    circulation of air and/or water

    sound and music

    transpiration

    magnetism

    current electricity

    etc.

The glass-jar garden example utilized in the Process Approach can be expanded beyond the accumulation of ideas by any or all of the three approaches described. A consideration of some of the above variables could generate ideas for experimentation. Some examples are:

> Is light necessary for plant growth?
>
> Do plants grow toward light?
>
> Do plants in sunshine give off oxygen?
>
> Is there a preferred color of light to enhance plant growth?
>
> How does temperature affect growth?
>
> How does air affect germination?
>
> How does humidity affect plant growth?
>
> Do plants react to sound?
>
> How do plants react to electrical stimulation?
>
> How much water does a plant lose?
>
> What is the lifting power of transpiration?
>
> Etc. etc. etc.

Seemingly this idea generation could go on forever. It need not. If you feel you have reached this point, you undoubtedly have reached the objective of becoming a generator rather than a duplicator. A teacher armed with sixty ideas is better prepared to teach than one armed with six ideas. And, one armed with six ideas is better prepared to teach than a teacher with just one idea.

When "sciencing," help children to:

- make complete observations
- distinguish observations from inferences
- look for the existence of patterns
- construct, when applicable, three-dimensional models to explain observed phenomena
- make simple drawings of problem situations
- make reasonable inferences and if sufficient data exist, make predictions
- state a hypothesis which could explain what happened
- raise provocative questions as opposed to trivial questions
- construct and respond to questions which tests a hypothesis
- state a problem in the simplest terms possible
- distinguish reasonable from unreasonable statements
- solve an analogous, simpler problem
- break the problem down into simpler more readily solvable subsets of the original problem
- ask "What did I leave out? What am I not thinking of?"
- stand back from the problem and enlarge one's scope
- quesstamate and compare the quesstamate with the final answer
- state anticipated results of accepting and rejecting assumptions
- constantly review your goals in light of the original objectives

WHAT MIGHT A PROFILE OF A SCIENTIFICALLY LITERATE INDIVIDUAL LOOK LIKE?

The learner

- is a believer.

    Has faith in the processes that allow for problem solving through experimentation.

- likes it.

    Experiences a joy to discovery, inquiry, and the resolution of questions and problems.

- respects it.

    Knows the power imbedded in the experimental processes for searching out solutions to problems.

- is conversant in the language of science.

    Can communicate in the sciences being knowledgeable of the vocabulary, concepts, and structure of science.

- recognizes that success in science is a four letter word.

    Work!

- knows that science is a non-ending series of queries and quests.

    Is aware of the open-endedness of science.

- is awed by the enormity, diversity, and complexity of the

    natural world.

    Is mindful in more than a simplistic sense of the grandeur of the natural world.

- is aware of the universality of science.

    Knows that science has no boundries - national, cultural, or ethnic.

- knows the contributions of science and technology.

    Understands the difference between science and technology, the complementation of one to the other, and their contribution to society.

- realizes that one can effect change.

    Knows that being part of the cosmos is accompanied by a responsibility to the cosmos.

SOME PEOPLE WILL BELIEVE ANYTHING IF IT IS WHISPERED TO THEM.

-Pierre de Marivaux

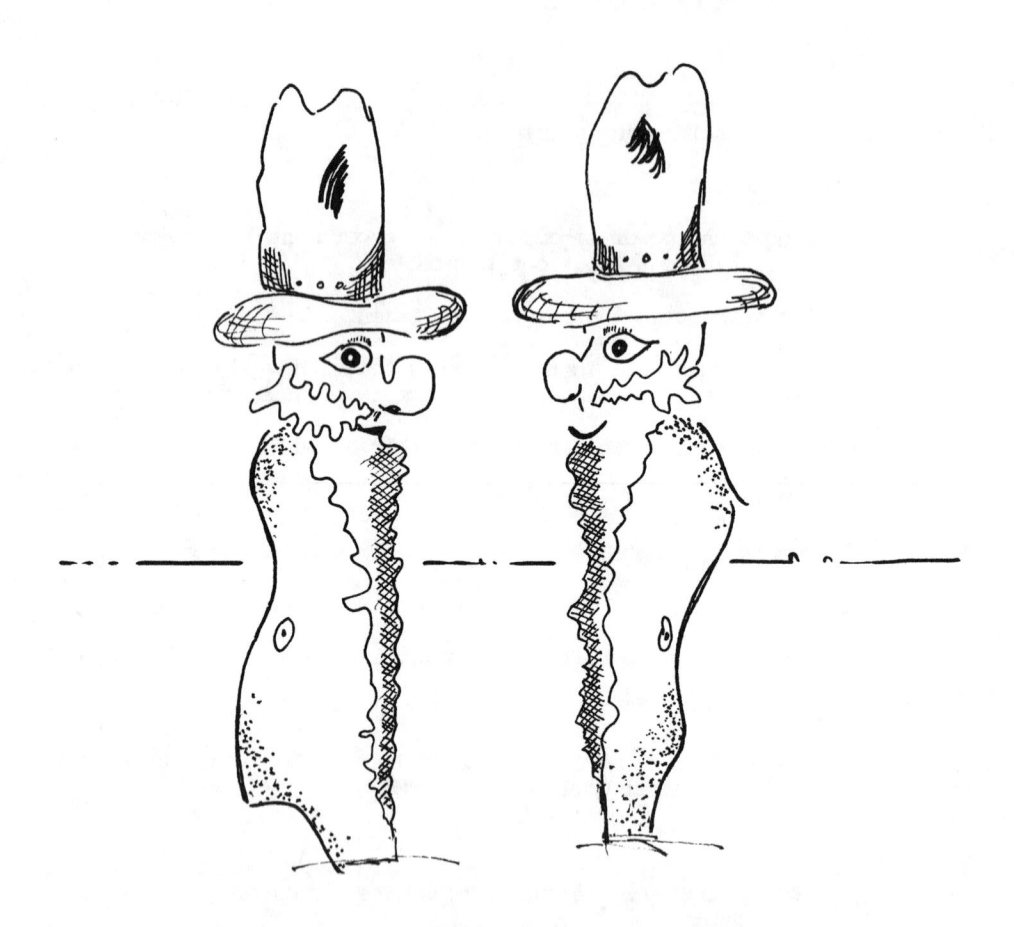

"SYZYGISTICALLY SPEAKING..."

Section III: Stimulating Thinking Through Science Instruction

CREATIVITY.

Our present time has to be labeled an exciting time. They say in fifteen years scientists will have discovered a drug that will make you smarter, live longer, change the sex of an unborn child, provide artificial eyesight for the blind, and allow for artificial growth of new limbs; ultrasonics will enable the control of pain and detect brain tumors; electric impulses will heal bone fractures; and it is even anticipated that we will have a substitute for blood.

Along with these GRAND ideas we need many _less_ grand ideas such as:

- A pair of reversible socks that are different colors when reversed. One pair would go with two different outfits.
- Parking meters that give change.
- A flourescent toilet seat eliminating the need to turn on a bright light in the middle of the night.
- Three way bulbs that do not have to be turned through all three stages.
- A chewing gum that can also serve as dental floss.
- Paint that changes color with differences in temperature, absorbing heat in the winter and reflecting heat in the summer.

Who thinks about these things? People! People like you and me. People who ask questions, some that border on the bizarre, for example:

- Is infinity like a mobius strip that winds back on itself?
- If everything is relative, what if anything matters?

. If mass is conserved, does this mean the universe is finite?

If necessity is the Mother of invention, then curiosity and intellectual engagements, resulting in diverse solutions to problems, may well be the Father of creativity.

What is creativity?

Creativity is a thinking process. It is any thinking process which solves a problem or answers a question in an original and useful way. Life itself is creative. No two people in the entire world are alike. And, more amazingly never again in all of time will there ever be another person exactly like any one of us. Each of us is a unique creation.

Creativity is a vacillating process where one oscillates from general to specific details and back again. It is a matter of changes in one's perspective ranging from macro-vision to micro-vision views of things. It is the process of observing in the common everyday things, uncommon things, so much so that the process of observing uncommon things becomes a very common event. If one dwells on the common, mundane every day things and never revels in uncommon views, it is not because one is uncreative, but rather because one has likely smothered creativity.

Creativity favors the prepared mind. It is being constantly alert to the unexpected. Expecting the unexpected without being aware of what to expect is a prime pre-requisite to creative idea generation. One rarely observes what one does not expect to observe. The greatest frontier is in the minds of man armed with a notebook, a handlens, curiosity, plus creativity.

Who is creative?

Creativity is something everyone possesses in varying degrees; everyone is born with the seed of creativity within one's self. Creativity once viewed as an extraordinary act, something ethereal, something reserved for a few, is now generally accepted as a common act. Creativity is a style of thought and action that is fundamental to the learning process. Creativity may well be the missing link in the learning hierarchy. Creativity is not acquired. One already has it to one degree or another. It can be nurtured. Its development depends upon the environment into which it is introduced and circumstances that condition it.

The Components of Creativity

Creativity is both process and product. There may be as many creative processes as there are creative people. The creative act is a function of knowledge, imagination, and experience. Knowledge is a necessary but not a sufficient condition of the creative act. Imagination is the ability to see what is, and to anticipate what is not. This may require repeated observations accompanied by a manipulation of knowledge by combining and rearranging facts into new patterns. The more experiences one possesses, the more new relationships one concocts as the manipulation takes place.

Recognizing the variety of processes associated with creativity, one general model appears to evolve. This is an engagement/resolution model. The first step in this model is the recognition of a problem, a statement of the problem, and the engagement with the problem. Once the problem is defined, the next step is the clarification and assessment of that which composes the problem. This engagement involves an analysis of the problem and various searches for a strategy that leads to a solution. This may be a tedious and sometimes frustrating step in the process. This frustration period may involve a temporary stalemate. Repeated ideas may be tried. Failures may ensue. Interest may wane as the solution appears to be even more remote. An impasse may be reached. Some individuals give up.

Some individuals persist. Those who persist and continue their search for a solution usually reach the "Eureka" or the revelation component of the creative model. This attainment is usually accomplished through the incubation of ideas and the translation of these ideas into active manipulations wherein a solution or resolution is reached.

Attributes of Creative Persons

Creative individuals tolerate uncertainty. They have the innate ability to be at ease with the strange, the bizarre, the mysterious, and the doubtful without any need to always search after explanations, facts, or reason. They have an incessant desire to create order out of chaos. And, they have the courage to persist to create that order to suit their own terms. This makes them often difficult, unconventional, and taxing to those around them. Creative individuals are more concerned with satisfying themselves in terms of self esteem then in pleasing others. Creative persons usually:

- Are dedicated towards their work and prefer to excel in their efforts.
- Enjoy variety in all engagements.
- Resist supervision. Are impatient to get on with things.
- Are critical of current conditions and seek alternative solutions.
- Immerse themselves in projects.
- Possess an abundance of energy and apply it in a disciplined manner.
- Are willing to take a chance, take risks, go that step beyond, risk security to explore new and challenging areas.
- Extend their thinking beyond conventional boundries.

- Have a strong self concept. Feed on recognition.
- Enjoy thinking and accepting challenges.
- Are persistent in the search for solutions to problems.
- Enjoy kaleidoscoping and manipulating objects and ideas.
- Can entertain diverse considerations as directed toward the solution of problems.
- Are intuitive and have a high irritability factor to what does not appear "right."
- Are curious about their direction and orientation in relation to that which surrounds them.
- Are not overly concerned with the unusual or inconsistencies, but instead search these out.
- Possess the perpetual curiosity of a child.
- Are both childish and mature vascillating from one to the other using assets of each to arrive at critical decisions.
- Prefer to be non-conformist. They differentiate blind conformity from deliberate or purposeful conformity.
- Are very well disciplined.
- Work very hard. Success comes from sustained work accompanied by extensive training and a perserverance in their area of interest.

The Divergent-Convergent Model

Inherent to success in creative thinking is the ability to capitalize on the assets of divergent thinking. Divergent thinking is the spewing out of many ideas from a basic or common source. For example, assume that you have a coin in your pocket. How many uses can you think of for this coin? You might have responded by stating:

- spend it
- drill a hole in it and use it as a washer
- drill four holes (or two) and use it as a button
- use it as a weight (fishing, paper, etc.)
- use it as a shim to raise something to make it level
- as a small wheel
- as a conductor of electricity
- as a piece of jewelry
- as some sort of spinning device
- hammer it into a flat sheet
- as a screw driver
- etc., etc., etc.

Originality springs from divergent thinking. Each added idea expands on the initial thought bearing forth a cascade of ideas.

By contrast, one converges in convergent thinking. Convergent thinking is directed toward one focal point; one response; one answer. An example might be: "I have something (a coin) in my hand. What is it?" Responses, thus requested, are focused on one, correct answer. This invites guessing until someone states that the item is a coin.

Neither convergent or divergent thinking exist in isolation. These two processes complement each other and are part of a cyclic, Divergent-Convergent model.

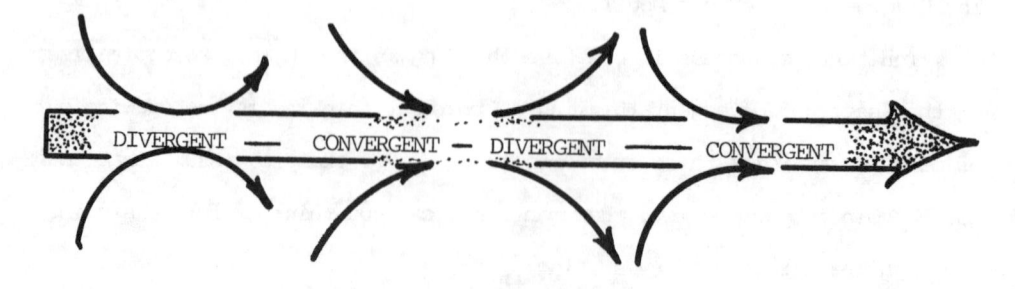

DIVERGENT — CONVERGENT — DIVERGENT — CONVERGENT

In order to be useful divergent thinking must expand thoughts by providing a plethora of ideas. The practicality of this expansion necessitates some priority of selection towards the solution of a problem. The Divergent-Convergent model is a process of expansion, selection, testing of selection, and then, perhaps reverting back to the divergent process for a new level of expansion, selection, and testing of selection moving to a new level of expansion. The process repeats itself.

Science is a "natural" for creativity. The scientist is one of the most creative persons in our society -- frontiers in science are constantly being advanced by his creations. The qualities or characteristics of scientists are in many ways analogous to those traits associated with creative individuals. Scientists are curious, independent, and have an adventurous spirit for searching out the unknown. Scientists have been classified as risk-takers because they search for that solitary chance that an unmarked avenue of investigation might contribute to the solution of a problem. Scientists are further recognized as individuals with strong imaginations and inventive behaviors.

In science teaching, opportunities exist for the development of creativity in problem solving, experimentation, and model building. Science teachers who work at the development and transfer of creativity find excitement in the process and invariably generate new ideas for children and themselves.

How to improve your creativity

Creativity is something everyone possesses. Everyone is born with some creative potential. Our function as teachers is not so much to uncover why some people are more creative than others, but rather to find out how we can make more of us use and develop the creativity we already possess.

Creativity can be developed by:
- Applying one's self. Success in developing one's creativity is not a matter of luck, but rather a disciplined, enduring effort to achieve creativity. There are few short cuts. Enlightenment comes from persistent application. Every action should be thought of as a challenge to your creative talents.
- Getting involved with creative, productive endeavors. Throw your weight into something that really interests you. Get immersed in the endeavor. Accept the challenge of new involvements.
- Working and reflecting in an unorthodox manner. If you are doing something now in the same manner over the past three to five years, what with all the advances in our society, there probably is now a preferred way to do it. Brainstorm into tomorrow, don't sit on yesterday.
- Associating with creative individuals and creative projects. Creativity is infectious.
- Maintaining a positive self image towards the creative generation of ideas. Vitality and enthusiasm lubricate the creative process. Creativity is not fragile, but creative vitality is.
- Believing in your creative potential.
- Stimulating both the analytical and the intuitive hemispheres of the mind capitalizing on the advantages of both.
- Developing a plan of action for creative improvement, then carrying out the plan.

In the development of the creative process remember that:
- A prepared mind is a common factor in all discovery.

- Creative individuals exhibit a high element of serendipity and sensitivity.
- Creative individuals are problem-oriented, not method-oriented.
- Creative minds are unorthodox and independent.
- Creativity is, perhaps, our least tapped potential.

. . .

> The most incomprehensible thing about the world is that it is comprehensible.
>
> Albert Einstein

. . .

Barriers to Creative Teaching

- over planning

    While it is a recommended teaching technique to plan for more instructive material than one can possibly cover in an assigned period of time, it is desirable to allow time for those creative moments that take place in the classroom. What we uncover can be as valuable as what we cover.

- excessive conformity

    Everyone needs some level of conformity in dealing with everyday chores of life. Conformity facilitates many operations and reduces mundane tasks to simple, not-necessary-to-think-about efforts. A modest amount of conformity is necessary. Driving a car is a classic example of what could be termed "necessary" conformity. Creativity, while driving, could prove hazardous to one's health. Excessive conformity is conformity whose inclusion is not necessary, but yet it exists. When choices avail themselves to you, decide whether conformity is necessary or not. If not, exercise your creative right and be a non-conformist.

. same plan for all children

Everyone recognizes that each of us is different; and yet, to nurture individual appetites, we serve the same educational menu to all children.

. excessive use of stereotyped questions

Eliminate questions that call for simple yes-or-no responses. Avoid asking questions that invite guessing responses. Ask divergent questions as well as convergent questions. Match your questions to the full range of Bloom's taxonomy. When applicable, use open-ended questions as opposed to closed questions. Good questioning skill acquisition leads to problem sensitivity.

. overuse of textbooks

Textbooks can be supplemented by the use of daily newspapers, libraries, museums, and personal or community resources.

. emotional considerations

The most obvious barrier to creative teaching is the mind set of the teacher. Creativity is stifled when fear takes over. Being creative requires an adventuresome spirit. For those who are timid creativity becomes elusive.

Fear becomes a dominant force in retarding creativity; fear of failure, fear of making a fool of one's self, fear of relinquishing control of the class to the creative process, fear of taking a chance, or fear of maintaining a creative level.

Creative teachers are positive thinking individuals. They possess a strong, self image and they recognize their potential for greatness. They develop a curiosity about everything. They observe with purpose. They are thinker-uppers. They dream and fantasize but always with a thought on the reality of things. They search for the "best" in everything. They are "yes" people. They have high expectations of themselves and of others. They question freely. They use time wisely. They think a great deal.

It takes courage to chart new waters. Creative teachers display an openness to new experiences. They are risk takers. They dream; and then they engineer their dreams down to the "real" world. They are educational engineers designing and implementing creative ideas for children.

Sciences for the gifted

Gifted children know a great deal about science but they do not necessarily do much sciencing. Science in the primary years of elementary schooling is not generally a basic subject taught with regularity. Also, it is probably that portion of the elementary curriculum that teachers feel the least comfortable teaching. As a result gifted children in their primary years are provided few opportunities to engage in what could afford them the most joy and intellectual stimulation.

Gifted children through early accomplishments in reading are interested in a host of topics. From this broad range of interest several major areas of particular interest emerge. These are mathematics, history, and science. Because of the combination of history and science, a large majority of gifted children find historical geology to be particularly appealing.

Gifted children should acquire from a study of science such attributes as the questioning of all things, and the consideration of premises, contingent variables, and the consequences of proposed action. Subsequently, they should also acquire an attitude of demonstrative verification of data; a search for data and their meaning; and an admiration for logic and the inquiry approach as applied to the resolution of problems.

Who are the gifted?

Gifted children are those who by virtue of outstanding abilities are capable of high performance. These children include those with demonstrated achievement and/or potential ability in a variety of areas such as intellectual ability, leadership ability, creative or productive thinking, or the visual or performing arts.

Some obvious characteristics of the intellectually gifted are:
- an early fascination with reading
- an impressive memory for their age level
- a curiosity as to the "why" of everything
- a strong desire to search out problems, an enjoyment with puzzles, and seeking out complex mental engagements
- a willingness to accept challenges and a desire to assemble and disassemble things to see how they operate
- an awareness to inner thoughts and feelings
- a desire to reach out for mature advice and associations

Science tailored for the gifted

There is no segment of science reserved for any particular group of children classified by abilities. There is no segment of science reserved just for the gifted. Science is science. There is, however, science tailored for the gifted. This tailoring is a function of teachers and instructional developers.

Gifted children in elementary science instruction need special programs specifically designed for them. Gifted children are anxious to get to the heart of problems. They usually exhibit an impatience with the minutia of science and quickly want to be engaged with significant aspects of the problem. They enjoy project work on an independent or cooperative basis. They seek out the challenge involved with all engagements. They prefer to work on problems whose solutions can make a difference.

When preparing science lessons for your entire class stay with the basic lesson, and from this branch up and out to accomodate the gifted. The science does not change, just the instructional methodology. Most science involvements can be suitably adjusted from one level to another by your insistence on more rigor in the observing and the quantification aspects of the investigation. For example, with slow children, perhaps, it might be advisable to stay with simple whole numbers. Gifted children would be expected to refine the quantification and perhaps carry it out to whole numbers plus either common, fractional parts or decimal fractional parts.

Additional pieces of equipment, the mathematical language, and the reading requirements that you impose on a specific lesson can move the lesson from simple, broad interpretations of one's observations to sophisticated interpretations. The added use of manipulatory equipment such as timers, balances, thermometers, microscopes, etc., plus the level of questions you interject into the lesson increase gifted children's involvement in science in a more challenging manner.

REALLY WE CREATE NOTHING. WE MERELY PLAGARIZE NATURE.

- Jean Baitaillon

CHALLENGES FOR CHILDREN

- Design something that will remain in motion for three minutes.
- Design a new child's toy or game.
- Examine a product and trace the steps that are involved in making that product, for example, a paper bag, a cardboard box, or a flow-through tea bag.
- Take a door lock and key assembly apart and figure out how the key operates to open the lock.
- How is the material in a tube of striped toothpaste arranged so that the stripe comes out as it does?
- Take a watch or clock apart, reassemble this and keep it working.
- Using a series of gears and a miniature motor powered by household batteries, construct a device to accomplish some work-saving benefits.
- Attempt to determine how a push-button telephone operates to successfully reach a dialed, phone number.
- Design an automatic goldfish feeder.

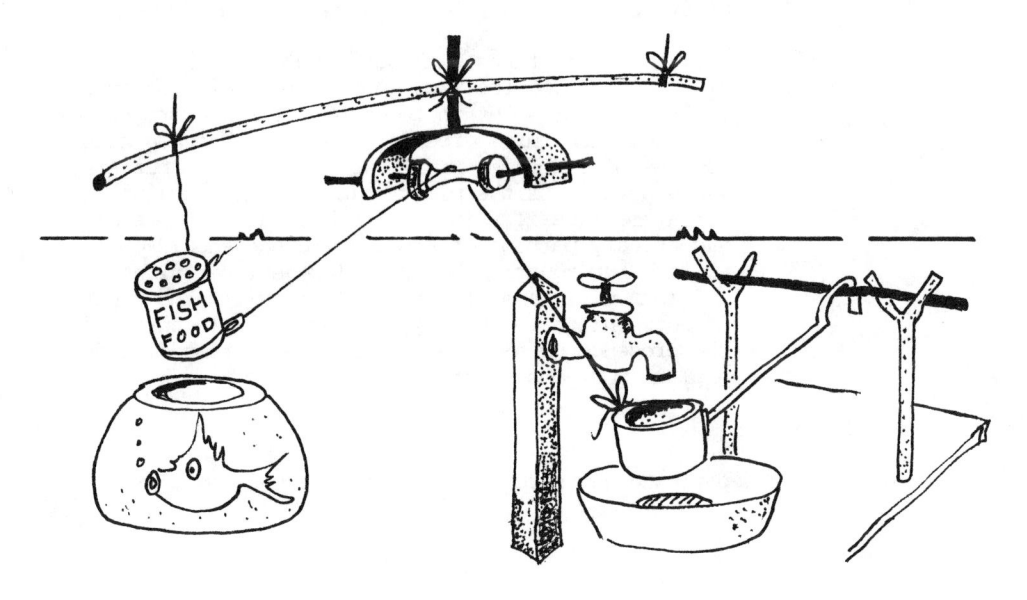

EXAMPLE OF A BASIC SCIENCE LESSON EXTENDED TO ACCOMODATE THE GIFTED

Basic Science Lesson: Plant Growth

* Science Activities for the entire class to accompany the text or other prescribed instructional materials

   Baggie Garden
   Tumbler Garden
   Construction of Terrariums

       Plus: Reaction to light, water, temperature, etc.

* Suggested Search Extensions to Accomodate the Gifted in Science

   Plants and Geotrophism
   Plants reacting to:

      sound
      electrical stimulation
      magnetism
      pressure
      various soil types
      varying degrees of salinity
      varying amounts of fertilization
      insect infestation

   Development of the "S" curve model for plant growth.

Suggested Interactive search

   Fill four test tubes with aquarium water to within one or two inches of the top. Add 3 drops of bromthymol blue indicator solution to each test tube (bromthymol blue solution turns green if the water becomes acid because of carbon dioxide). In test tube 1 put a leafy portion of the plant Elodea: in test tube 2 put a small snail: in test tube 3 put both a snail and a leafy portion of Elodea: tube 4 should contain nothing but aquarium water. Place stopper tightly on all four vials and set them in strong light but not direct sunlight. Observe any changes. In terms of the presence or absence of dissolved carbon dioxide explain your observation.

Provocative Question

   Does a plant experience any shock when portions are snipped off?

REFERENCES FOR THE LESSON "PLANT GROWTH"

Bibliography

 Budlong, Ware T., <u>Performing Plants</u>, Simon & Schuster, N. Y., 1969
 Selsam, Millicent E., <u>Plants that Heal</u>, William Morrow & Co.,
  N. Y., 1959
 Wickers, David & John Tuey, <u>How to Make Things Grow</u>, Van Nostrand
  Reinhold Co., N. Y., 1972
 Biological Sciences Curriculum Study, (BSCS) # 18184-6 Plant
  Growth & Development Laboratory Blocks
 BSCS, Pamphlet Series, Plant Biology
 BSCS, Collegiate Minicourse Program, Plant Structure and Education
  # 6374-5
 Judson, Horace Freeland, <u>The Search for Solutions</u>, "Modeling,"
  Pg. 111-130, Holt, Rinehart & Winston, N. Y., 1980

<u>Current Available, Portfolio of Previously Collected Newspaper and
Journal Articles Referring to Plant Growth</u>

<u>Scientific American Reprints Relative to Plant Growth</u>

 The Mystery of Corn - Paul C. Mangelsdorf
 Plant Growth Substances - Frank B. Salisbury
 Light and Plant Development - W. L. Butler & R. J. Downs
 What makes Leaves Fall? - William P. Jacobs
 Electricity in Plants - Bruce Scott
 Heat Transfer in Plants - David Gates

STIMULATING THINKING THROUGH SCIENCE INSTRUCTION

Science, more than any other area of the curriculum, has more excellent materials available for instruction than any other area of the curriculum. Major, nationally-funded curriculum projects for elementary instruction, such as, Science - A Process Approach, Science Curriculum Improvement Study, Elementary Science Study, other projects too numerous to mention, plus major individual efforts, make available to us an almost unlimited supply of creative, innovative ideas to stimulate thinking. It is not within the scope of this publication to delineate these programs. The single most comprehensive source for determining what is available in science curricular development is the International Clearinghouse on Science and Mathematics Curricular Developments. The University of Maryland's Science Teaching Center in College Park, Maryland has taken on the task of summarizing and publishing the majority of on-going projects in science and mathematics curricular development. This compendium of programs lists titles and subject areas covered and summarizes the goals and characteristics of each project as well as the evaluation techniques utilized. This publications contains approximately 400-500 pages and lists over 300 projects.

Learning is incremental. A program to stimulate thinking is an on-going process and should commence in the early grades. Instruction to stimulate thinking can be represented by a three-tiered model. This model represents simple to complex levels of thinking.

The entrance level of the three-tiered model is that level where the teacher is the dominant figure, leading discussions, posing problems, instigating investigations, providing clues, asking provocative questions to spark interest, providing materials, and effecting closure with the activities. These thinking engagements are usually activities whose solutions can be reached in a short period of time. These activities do not involve a great deal of equipment, finances, or preparation. They stimulate thinking in an easy, light provocative manner. This level sets the stage for the 2nd level by presenting numerous, high-interest encounters.

Some Level 1 involvement exercises are:

- If you could relocate your eyeballs where would you relocate them? Why?
    * Variable answer
- A man rode into town on Thursday. He stayed three days and left on Thursday. How might one explain this?
    * The horse's name was Thursday.
- How many 3¢ stamps in a dozen?
    * Twelve
- I have two coins. <u>One</u> is not a nickel. Together these two coins add up to 55¢. What are the two coins?
    * One coin is not a nickel but that doesn't mean the remaining coin can't be a nickel.
- How would it be possible for a person wearing pants to put one's right hand in one's left pocket; and at the same time put one's left hand in one's right pocket?
    * Put them on backwards.

Level 2 is an intermediate level for the promotion of thinking. At this level the teacher abdicates some authority. The teacher, at this level, promotes student initiative in the searching out of problems, stating questions, identifying areas of interest, acquiring or substituting materials, and in effecting closure. At this level the problems are more involved. A more sustained effort is needed to arrive at the solution to the various problems.

Some Level 2 involvement activities are:

- How does one know that the light in the refrigerator really goes out when the door is closed? Devoid of climbing inside the refrigerator or drilling a hole into the refrigerator, design an investigation to establish that indeed the light does go out.
- What is your decision? By some unusual circumstances, a small airplane crash-landed on a large ice floe in the upper North Atlantic Ocean. All in the small plane survived. The group consisted of the pilot, who was a world famous explorer; a movie star; a minister; a young child; a pregnant woman; an old man dying of a terminal illness; the Secretary-General of the United Nations; and a famous eye surgeon, who was on the brink of an important breakthrough in his research which might aid thousands of blind individuals to see. The ice floe is drifting southward into warmer water and melting rapidly. It will melt completely before it reaches any major shipping lanes. Enough flotation equipment was salvaged from the airplane to construct a small raft for only one person. When the ice floe is completely melted, which individual should be allowed to escape using the raft? Or should no one use it? What is the basis for your decision?*
- You have six spheres. They are of equal size and volume. One of the spheres has a mass different than the remaining five

---

*From Alfred De Vito and Gerald H. Krockover, <u>Creative Sciencing - Ideas and Activities for Teachers and Children</u>, 2nd Edition, Little, Brown and Co., Boston, MA., pg. 90-91.

spheres which individually all have the same mass. Using an equal arm balance; and only being able to make two separate weighs, isolate or identify the sphere with the different mass.

- The Cornflake problem: Tommy was the first one up in his house. He decided to prepare his own breakfast. He filled a bowl nearly to the top with crispy cornflakes. He added milk. Then, he cut a large banana into slices. These he placed on top of the cornflakes. This brought the milk level up to a little more than one half of the bowl. The telephone rang in another room. Tommy answered it and talked for twenty-five minutes. When he came back to the kitchen, Tommy noticed that milk was spilled all around the bowl. What explanation might you give for Tommy's observation? Some additional facts are: No one in the household came near the bowl filled with cornflakes. There are no pets in the house. Tommy left his spoon in the cereal bowl. He added one spoon of sugar to the cereal. What do you think happened?

- Given a sheet of paper 8½" x 11" fold this into a cylinder which has a base with a circumference of 8½" and a height of 11". Using the same sheet of paper, refold this so that the new cylinder has a base with a circumference of 11" and the height is 8½". Do the two cylinders have equal volumes or unequal volumes? Check and verify your statement.

Level 3 is characterized as the highest level of the three-tiered model. It is at this level that the teacher removes oneself from the role and becomes more a broker, arranger, or facilitator of learning assisting only when requested or as needed to keep things within prudent

137

bounds. This level is referred to as the project level -- projects based on the student's own interest. Project work may be an independent endeavor, however, it is not unusual to observe work accomplished in small groups. Achievements at this level usually culminates in some realistic goals with some end product or some societal impact generated from the action.

Some Level 3 examples are:

- If you were assigned the task, how might one organize a Saint Patrick's day parade being held in a large city?
- How would one have engineered the building of Stonehenge if you were there at that period of time in which it was built?
- Your favorite fast food restaurant wants to build an outlet (or another one) in your hometown, where would be the best location?
- Based on water, mineral, soil, energy, geography, etc., how might the 50 states of the United States of America be changed to best serve all the people of the United States?
- Redesign your school building so that it better serves your needs. Defend your suggestions.

CREATIVITY AND QUESTIONING

> A good question is a problem half solved.
>
> De Vito

Good questions lead to good responses. Good questions are stepping stones directed towards the solution of problems.

Cognitive learning is the accumulation of content knowledge and may be classified in six levels: knowledge, comprehension, application, analysis, synthesis, and evaluation.[*]

The KNOWLEDGE level is defined as that level of cognitive learning where recall of information is dominant. This is the lowest level of learning. It is the easiest level to assess.

COMPREHENSION, the second level of cognitive learning is concerned with understanding. Does one understand what one knows? Does one comprehend? In your own words, can you communicate your understanding of previously acquired knowledge?

APPLICATION, the third level of cognitive learning means to apply what you know, and comprehend, towards the solution of problems. It is one thing to know, it is another thing to understand what you know. And, it is another thing to apply what you know. The act of application, while a higher level of accomplishment, is contingent upon the two subordinate levels of knowledge and comprehension.

ANALYSIS is that level of cognitive learning concerned with the ability to separate things into their component parts for close and

---

[*] Benjamin S. Bloom, ed., *Taxonomy of Educational Objectives, Handbook I, Cognitive Domain* (New York: McKay, 1956)

critical analysis.

SYNTHESIS in the cognitive learning hierarchy is that level pertaining to the uniting of separate parts to make a new form and to encourage divergent thinking.

EVALUATION is the capstone of cognitive learning. It is the most complex level. Its use necessitates combinations of the other five levels.

The scope of the six levels of cognitive learning are used to hierarchially structure learning. These levels are also applicable in classroom dialogues, involving thinking, writing performance objectives, and in testing (evaluation). In all three areas, it is suggested that teachers consider all six levels and not restrict themselves to one or two levels. These six levels can be applied to questions asked in classroom instruction. When asking questions, include questions from each level ranging from knowledge to evaluation. When writing performance objectives, include stated objectives that are representative of the entire spectrum of the cognitive learning levels. Also, when testing spread your test questions across the entire cognitive scale. This diversity in questioning, in stated performance objectives, and in testing will serve a broader range of children in a more satisfactory manner.

. . .

What everyone knows isn't worth anything. Ask a provocative question.

De Vito

. . .

The value of question asking to the learning process has long been recognized. Unfortunately, most children in classroom situations ask trivial questions such as, "In pencil or ink? Does spelling count? Will this be collected? Will this be graded? Will we get extra time to work on this?" And, on it goes. These questions are, as are most of the questions asked in elementary school classrooms, clerical in nature. For children, these are classroom survival questions. But, they contribute little to the quality of the cognitive learning process.

The art of questioning is an on-going act. Proficiency continues with improved practice. One gets better and better at it. Improved questioning can arise from:

- avoiding the overuse of low-level cognitive questions.
- asking a better mix of questions as outlined by Bloom's taxonomy.
- becoming a better listener of children's questions and statements.
- avoiding questions that can be answered by a simple, yes-or-no response.
- initiating more provocative thought questions.
- stating questions succinctly.
- the avoidance of answering one's own questions.
- eliminating guessing-response questions.
- fostering questions commensurate with children's abilities and past experiences.
- promoting, when possible, open-ended questions as opposed to closed questions.

Concomitant with better questioning acquisition by teachers, children should be trained to seek the same end. Good questioning technique, by example, is not enough. Children need to become involved question askers. Opportunities can be provided in a variety of ways, such as:

- Start each lesson by asking children to state, "What questions do you want answered relative to the topic under discussion?"
- Don't ask, "Are there any questions?" Better to state "What questions are there?"
- Permit a periodic question and answer session.
- Initiate question boxes; weekly, question-asking bulletin boards; and best question of the week awards.
- Play question games, for example, a young boy passes an empty, haunted house. Its window panes are knocked out. He throws a ping-pong ball through one of the broken windows. It disappears into the house, but is swiftly returned, flying out from the same window. How might one explain this? What questions would one like to have answered in order to more accurately account for the observed phenomenon?

A Question Asking Situation: Hathaway Attawhey Went That-a-Way

A crime has been committed. You are called in as the best detective available.

An old prospector was killed. He had been badly beaten and his pockets were turned inside out. The death of the prospector was established as having been caused by a blow to the head. The body was found near his diggings where he was checking out a gold claim in the lost Apache Mine Area of Arizona. This area was located within the sacred Indian burial grounds forbidden to all outsiders. The prospector's partner, Hathaway Attawhey was missing. The old prospector's mule was found standing next to the body with blood on his front left foot. Hathaway Attawhey's personal things were still in camp. From all indications the food supply was low.

What happened? Who did what? Make some inferences.

Assuming that you are there list as many observations, inferences, and hypotheses which you might make relative to this situation. What questions if you had the answers to, would assist you in solving the crime?

"What would happen if..."

Subsequent to the act of encouraging good questioning is the inclusion of the creative act of the "What-would-happen-if" technique. The inclusion of this technique provides open-ended avenues for mental explorations. Some examples are:

- What would happen if we had two left (or two right) hands, each hand oriented in similar position?
- How might things be different if we could cancel out the force of gravity whenever we so desired?
- How different would the world be if animals could talk?
- What would change if we never felt pain?
- What would happen if humans had gills enabling them to live both in and out of water?
- What might happen if we could predict all future events?
- What might the world be like if all weapons of war were eliminated?
- How might things change if all our dimensions were doubled, for example, height, weight, width, etc.?
- What might the world be like if friction was eliminated?

- What might the world be like if humans did not have to eat to stay alive?
- How might the world be different if we did not have the metals of the earth available to us?
- What might the world be like if water when frozen into ice became heavier (more dense) rather than expanding and becoming lighter?
- What might the world be like if humans did not have eyes?
- How might the bow and arrow have evolved? The boomerang?
- What would the world be like if we could store energy?
- Does learning use or create energy? Explain.

<u>Ask provocative questions, for example:</u>

- Is nothing something?
- Where does the color white go when the snow melts?
- What is beyond the universe?
- Is hair color related to strength and thickness of the hair?
- Do all parts of a plant grow at a uniform rate?
- Do all parts of a leaf grow equally fast?
- How do cells of the body know what to become?
- Does grass hurt when one walks on it?
- What caused the dinosaurs to disappear?
- Are there really UFOs?
- What remains when we have forgotten all that we have been taught?
- Why must we all die?
- Why do quantities of salt, adequate to kill most other plants, not kill asparagus?
- Why does every geometric shape in the universe seem to tend

towards sphericity?
- How can some forms of life regenerate lost limbs, etc.?
- What allows insects to be so powerful for their size?
- How do pigeons find their way home?
- Why do some varieties of the deer family shed their horns each year?
- Are we all part of a universal cycle?
- Why don't animals cry?
- Do insects, bugs, etc. have fears?
- Why does a wet sponge soak up water more quickly than a dry sponge?
- Why can't one tickle oneself and laugh in the same manner as when someone else tickles us?

Additional Suggested Creative Involvements

- If you had to design an easy mass exit from the room you are in (eliminating the existing door), what would you propose?
- How could you tell a blind man what a shadow is?
- Explain how you could thread a needle or tie a shoe lace using only one hand.
- What puzzled you today? How might you resolve the puzzle?
- What did you invent today? What good ideas did you generate?
- Badminton, volley ball, and tennis all have something in common. Design a new game utilizing what is common to all three games and relate this to the game of basketball (give it a name); or if you prefer, design another game using another sport other than basketball.

- Redesign or alter the shape and use of something, for example, a watch; elaborate on where a watch could be worn to better advantage other than the wrist.

How and Why Questions

Why and How questions are questions of special concern. "How" questions deal with relationships and the processes of science. "Why" questions generate explanations. Each type has its place in the curriculum. "Why" questions are touted by teachers as those questions which engender "thinking" responses. However, "Why" questions require some caution in their use. "Why" questions can be difficult, and some times impossible, to answer, for example, "Why is there a universe? Why is grass green and not blue? or Why must every one die?" "Why" questions should be qualified to avoid complications in the various "stages and ages" of children's understanding.

Model Building -- THE FORGOTTEN PROCESS SKILL

> Model, model in my mind
> Do I see you there so kind?
>
> Conjured up in spirit bright.
> Serve me in my queried plight.
>
> Attend me deed you do.
> Til your usefulness is through.
>
> Return thee then to the cauldron's brew
> Til next to be fabricated anew.
>
> De Vito

Models exist and are used in the elementary school science curriculum. Many times they are used unknowingly and their full value is overlooked. Some common uses of models in the elementary science program are depicted as cycles (weather, erosion, rock, carbon, etc.), as graphic curves (the "S" growth curve, the normal curve for variations in height, weight, pulse rate, etc.), as pictorial representations (contour maps, graphs, etc.) and as abstract formulas (Newton's second law of motion, $F = ma$, Speed equals the distance traveled divided by elapsed time, $S = \frac{d}{t}$, etc.).

Models are anomalies in that they can serve the slow child as well as the gifted child in science. It seems they have something to offer everyone. Models assist us in understanding complex systems or events by describing these systems or events in simple terms. This is accomplished by replacing intricate and complex situations with simpler and more familiar analogies. The more adept students construct models. The less adept students benefit from difficult situations simplified and presented as models. If a picture is worth a thousand words, a model is worth ten thousand words.

Models come in a variety of forms. Also, they mean many things to many people, for example, "The model was modeling model clothes." This confusion probably stems from the fact that the word MODEL can be used as a noun, a verb, and an adjective. In reality the word model should not be confusing. Basically there are three types of models: <u>Concrete Models</u> which are replicas of the real thing such as a scaled down version of a train, a boat, or an airplane. Usually there is a one-to-one correspondence between reality and the model. <u>Analogue models</u> which are models that utilize the transformation of properties by converting or substituting one thing for another, for example, color or shading on a map used to convey elevation. And, <u>Abstract Models</u> which are models that express the properties of the thing represented in a symbolic manner, for example, a chemical equation or a mathematical formula.

Concrete and analogue models are easily understood by elementary children. However, abstract models which are more challenging and more useful are a bit more difficult to comprehend. An abstract model is an invention which appears to be appropriate for the understanding of some realm of data and which provides a plausible explanation of a phenomena with which a child is confronted. There must be a "fit" between the data and the model. A model is judged by its success in ordering data and in making verifiable predictions from the data. Models are the creative heart of science. They suggest ways of expanding the theory embedded in them. Models are idealized expressions and as such are apt to be less complicated than reality and thus easier to comprehend.

In working with models, analogies are often used to assist in understanding a phenomena. The idea that matter is composed of atoms can be represented by an analogy. The atomic model of matter represented by styrofoam balls and sticks is an analogy. The solar-system model wherein the sun is represented by a baseball with elliptical patterns drawn about it to depict the routes of various planets and marbles placed on these concentric ellipses to represent planets is also an analogy.

Models, if they are good models, can be used to predict. Can you possibly predict what other solar systems beyond our own might look like and how they might behave on the basis of models that you have seen of our solar system? Models are not necessarily true representations of that which they help us to understand and they should be discarded or improved upon as soon as they no longer serve the purpose of assisting in understanding a particular phenomena. If returning astronauts, and other space scientists, furnished us with new information about outer space which altered our current idea of space we would need to revise our model. We simply construct a new one and discard the old. Models are tested in terms of usefulness and hence are only retained as long as they can help us to understand and to make correct predictions.

Models are fun to design and fun to construct. Whether concrete, analogue, or abstract, models aid us in problem solving and thinking. They help us to predict when experiments are impossible, and, they suggest hypotheses worth testing.

The biggest danger in using models is that if we are not careful we will begin to think of a model as the real thing. This is like being wrong with confidence. Models invite over-generalizations. For example, a computer is no more a brain (though it may be a model for it) than a tractor is a muscle, or, as in our previous example, than a baseball is the sun. But sometimes we get so attached to the model that it clouds our vision of the real thing. The price of the use of models is eternal vigilance.

Another disadvantage in using models is that in our initial construction of the model we might include things that are unimportant and in no way affect the outcome. And also we may exclude some things that are important. Therefore, our predictions based on these inclusions or exclusions may be in error. Models involve dangers and may, at times, prevent the solution of a problem from being reached. It is too easy to assume that a model is a good model and that it does provide a sound analogy. In order to prevent this from happening, a model to be used for solving problems and other purposes should first be validated. This means that evidence should be collected to demonstrate that the model is a sound analogy and behaves in ways similar to the phenomenon it represents.

Mystery Containers

A mystery container is an opaque container into which an object is placed. The container is then sealed. Through external manipulations of the container, one, unfamiliar with the identify of the object, is to describe the object and if possible name it. More than one object may be placed in the container. The interior of the container may be altered to vary the path of motion of the object inside the container. They can be simple or complex in design. Success with children is predicated on their knowing how to "play" the game. Once they know what is expected of them, the game of mystery containers moves smoothly.

An introductory mystery-container exercise

Select a container, for example, a cigar box and any three objects. These could be three, wooden, geometric solids such as a cylinder, a sphere, and a cube.

Children should be involved in direct observations of all three objects. They should fully describe each object in terms of individual attributes. For example, all objects, providing they are made from the same variety of wood, are of similar hardness, color, density, texture, flotation, etc. To distinguish one object from the other, <u>differences</u> become more important than <u>similarities</u>. Each of the three, geometric solids has a unique property that allow that object to be separated from the other two objects. These properties are:

- The wooden cylinder in an upright position can slide. On its side, it can roll.
- The wooden sphere can roll in any orientation. And, it can roll in any direction.
- The wooden cube cannot roll in any orientation. But, it can slide in all orientations and in any direction.

At an early stage of introduction to mystery containers, it is desirable to stress the concept of similarities and differences. Having done this, now unbeknown to the participant, one of the originally described objects is placed within an opaque container such as a cigar box or shoe box. It is sealed. External manipulation of the container begins and subsequent observations are recorded. This approach provides a base of information from which children can make decisions. This strategy involves careful manipulation of the container to determine if it is clear of any obstruction. This may be accomplished by sweeping the internal object around the inside perimeter of the container; then, moving it across the entire container; then inverting the container repeating the sweeping process covering the entire internal surface area. Through continued use of this involvement, children refine their ability to interpret through indirect observations.

Once the child acquires a strategy of "What and How" to utilize mystery containers, more complex, mystery containers could be introduced.

Variations of mystery containers

- vary the object

    Select objects for placement into the container that by their attributes communicate something. A cotton puff placed inside a container communicates little. It does roll, but this may not always be communicated to an external observer. By contrast, a plastic egg with two, steel, ball bearings placed inside of it communicates more of a message than a cotton puff, a toothpick, or a rubber band. The plastic egg, rolls in an eccentric fashion. It makes more than one sound as it collides with an internal wall. It connotes a plastic sound when it strikes the sides of the container. And, it has substantial mass.

- Vary the interior of the container

    Vary the interior of the container by positioning obstructions within the container. Identify the internal object for the participants. Given the identify of the object, then, by their external observations, have the participants infer what the internal construction of the container might be.

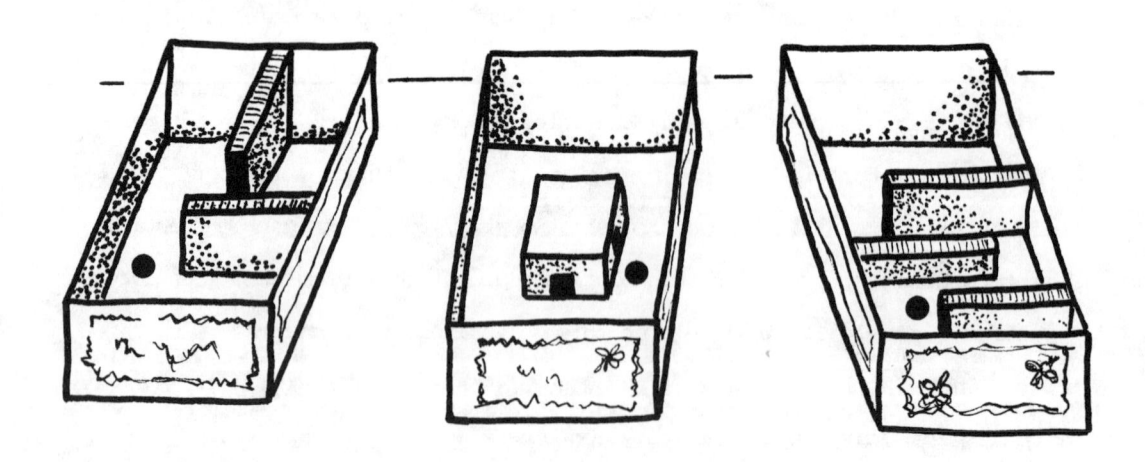

- vary the internal surface of the container

   The internal surface can be altered by gluing a sponge to one end of the container, by attaching sandpaper to another wall, by cementing corrugated cardboard to still another wall, and by positioning a magnet near or at the remaining wall. With the interior object identified, for example, a steel ball, the children through manipulation of the container, describe the altered, internal surfaces of the container and record any other indirect observations.

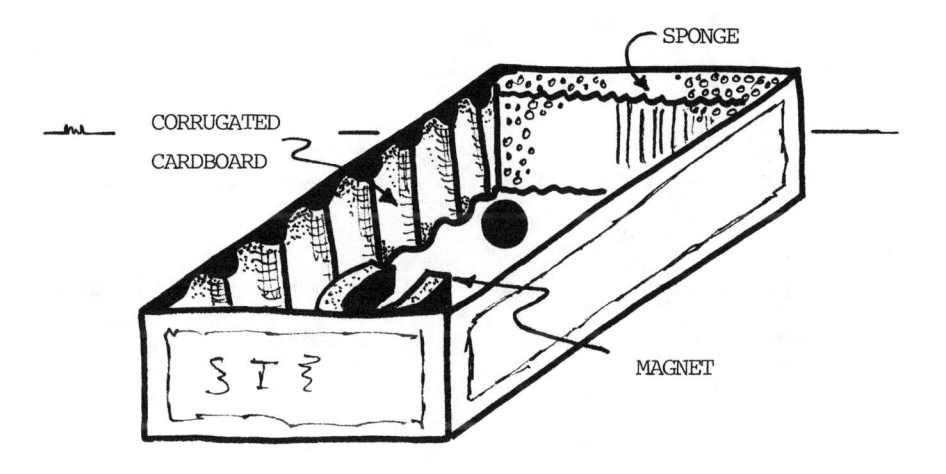

Mystery containers are highly creative. To maximize their contribution to the learning process, some precautions should be observed.

- In the use of mystery containers, variations in the internal construction or the object selected, care should be taken not to make them too simple or too complex. Beginners tend to make mystery containers to easy or too difficult. A steady progression of complexities enable children to continually refine their strategies for mastery of the skill of modeling in science. The only limitations on variations that can be used are one's imagination and the practicality of the variations as one moves from simple to complex levels.

- Time is a concern. The use of mystery containers is time-consuming. Some teachers express concern that too much time is wasted which might otherwise be devoted to learning the content of science rather than the process of model building. However, model building is a rehearsal for reality. And, it is a valid part of science.

- Noise can inhibit the use of mystery containers. Careful listening is necessary to adequately interpret the contents of a mystery container. Thus, one cannot instruct well using a large number of mystery containers in a classroom at one time. Several could be used if small group instruction is practiced.

- Mystery containers invite frustration. The mystery object's attributes are not always readily interpreted. Impasses are reached. Persistence is needed to change one's approach and to consider attacking the problem from a different perspective. In the upper-grade levels, mystery containers are rarely open to reveal the unidentified object. However, to keep the involvement alive, the teacher may periodically provide provocative clues. In the primary grades, children have short attention spans. And, they do not possess the ability to maintain interest over a sustained period of time. Thus, more often than not, teachers, at the appropriate time, reveal to the children the contents of the container.

Despite the shortcomings of mystery containers, they do provide children with opportunities to approximate certain patterns of behavior characteristic of scientists. These are: creativity, persistence, fastidiousness to details, open-mindedness, and knowledgeability.

Mystery containers can be expanded to operational mystery containers. These are containers wherein the contained object is eliminated and internal working parts with external reactions are substituted. This introduction challenges children to determine what is happening internally that is consistent with what they manipulate and observe externally. This variation of the mystery container provides children with an opportunity to use mind and hand to construct their own operative mystery containers as well as interpreting someone else's construction of an operative-mystery container.

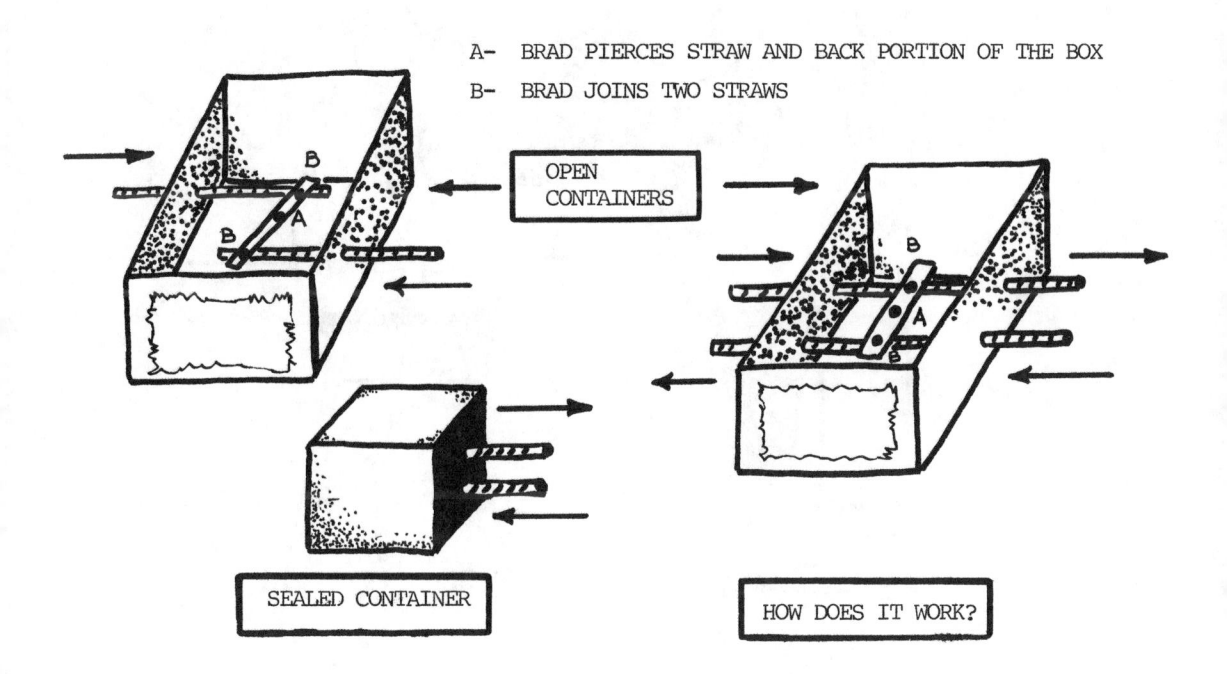

A- BRAD PIERCES STRAW AND BACK PORTION OF THE BOX
B- BRAD JOINS TWO STRAWS

OPEN CONTAINERS

SEALED CONTAINER

HOW DOES IT WORK?

Scientists deal with mystery containers all the time. They do not call them mystery containers, but they exist. Scientists faced with problems or dilemmas may never be able to "open" the situation and must, for at least the present, work indirectly. For example, a scientist cannot climb inside an atom, the scientist is in a sense dealing with a mystery container. By observation and experimentation the scientist creates models of what he thinks is going on inside the atom. As additional information is accumulated the model may be revised, but for the present, based on current knowledge, it is accepted as adequate.

Umbrellic Science Model

Science can be and is taught in a variety of ways. The most common instructional procedure exercised in elementary school classrooms is to lean heavily on a pre-selected or inherited science text and to embellish this with ancilliary materials. This common approach is somewhat cyclic in that it consists of <u>reading about</u>, <u>discussing it out</u>, and perhaps, <u>acting it out</u> through activities that are designed to confirm or reinforce the text material. Science instruction as it progresses in text form, from sub-chapters to chapters, and on to units is repetitive and conveys that there is a staccato rhythm to searching and understanding science. The organization of texts and the predetermined time allotment of public school teaching units almost mandates this interpretation of science. Science, however, is more a web than a sequenced staircase. Science should be represented in a variety of time spans. Science instruction may be portrayed as an umbrellic model rather than as separate, discrete entities disjointed one from the other.

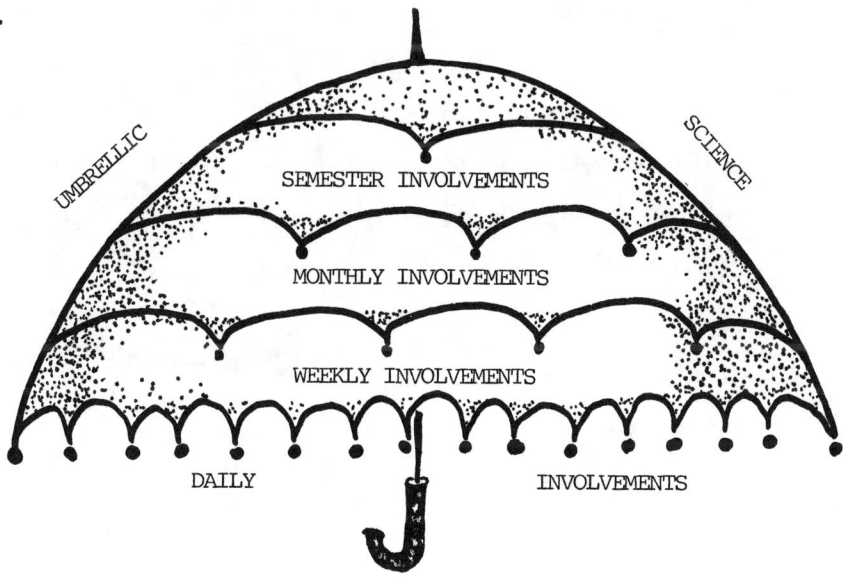

In elementary schools activities occupy most of the science involvements. Experimentation rarely occurs. This is due, in part, to the time constraints of a fixed period of science instruction and the feeling teachers have for "wrapping" things up. It is almost a rule, "Let there be no loose ends left for tomorrow." Some educators have labeled this phenomenon "Wrap-a-phobia."

Science should include along with daily, activity involvements, investigations that extend for a week or two; some that extend for a month or two; and experiments that extend over a semester or two. This mixing of activities and experiments conducted over varying time spans provides a more accurate picture of what science is and what science involves. Also, this diversity of involvements provides a broader spectrum of on-going science engagements which would appeal to gifted and talented children.

TIME-SPAN EXAMPLES:

Daily - short span of time, usually finished in one period of time

Topic: air pressure

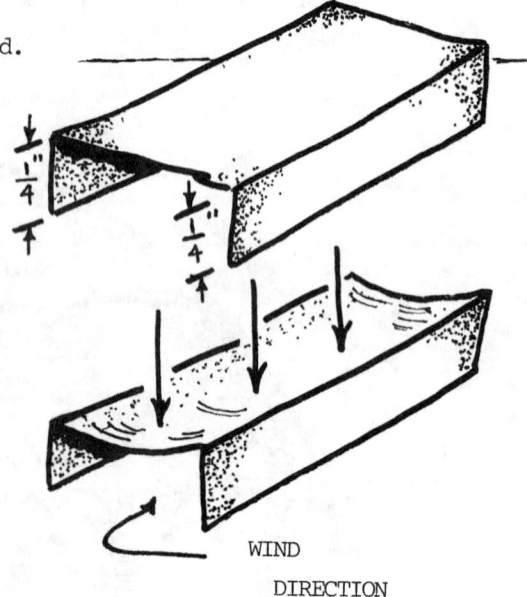

Each child is given a 3 x 5 card. The card is folded to form a bridge. Desk tops are cleared. Instructions are to place the card in the middle of the desk. The task is to use one's breath to flip the card over. The harder one blows, the more the card depresses itself. The task becomes impossible and one seems to work against one's self. A discussion ensues. Questions are asked, answered, and sometimes deferred. The bell rings. Science is over.

These daily sorties are a valuable part of the science program. However, they need to be augmented by:

Weekly - short span of time, usually extends over two or three periods of science instruction

Topic: crystal growing

Boil one quart of water. Add three ounces of powdered alum to the boiling water. Continue to let it boil for three to five seconds. Remove this from the stove. Pour the liquid through a filter into a jar. Let this cool for a twenty-four hour period, interrupted only by an ocassional tap to the jar to assist in the formation of the crystals.

This lesson weans the students away from the notion that science exists in fifty-minute packets.

Monthly - medium span of time, usually extends for several weeks, daily attention is required to keep it going.

Topic: growing seeds

Lima beans are planted in three different containers.

Three different types of soil are used - loam, sand, and

clay. All conditions are kept equal except the condition of the soil which is the manipulated variable. The responding variable will be the plant's growth over a period of time.

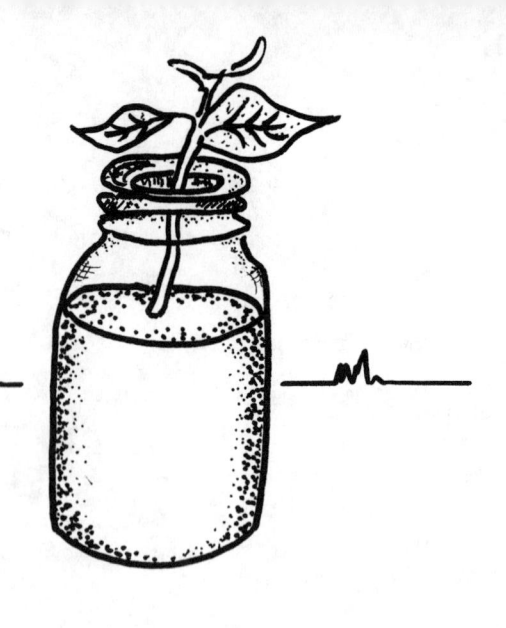

This involvement can go on concomitantly with daily classroom involvements such as the air-pressure activity, or the crystal activity. This involvement necessitates daily or weekly watering, recording of temperature, constructing temperature variation charts or graphs, measuring growth of the plant over time, recording observations of growth of particular parts of the plant, calculating mass increases, etc. Meanwhile the plant continues to grow. It may even bear fruit. And then, eventually it will die. This science involvement depicts science in a protracted form as opposed to fifty-minute, vignette, portrayals.

Semester - long term span of time, usually extends for months or possibly the entire school year.

Topic: the earth's rotation and revolution

This is an example of an on-going activity that provides an opportunity to collect and accumulate data. This data is collected over an extended period of time as one observes changes of a stationary object's shadow, for example, a flag pole. Shadows can be recorded daily at the same time each day over an extended period of time; or observations can be made to note shadow changes for a period of one day, taking readings at hourly intervals. Recorded data are interpreted and conclusions drawn. Astronomical observations also provide long-term investigations.

Long term span-of-time activities are particularly applicable to the development of models to explain specific areas of science. These models give warp and woof to the weave of science. These models can be developed following various themes. One such model for the process of dissolving is:

A Model For Dissolving

STEP A. RELATIONSHIP OF SURFACE AREA TO THE RATE OF DISSOLVING

STEP B. DISSOLVING AS INFLUENCED BY DIFFERENCES IN TEMPERATURE, LIQUID, AND MOTION

STEP C. PHASE CHANGES

STEP D. SUBLIMATION

STEP E. SATURATED - SUPERSATURATED SOLUTIONS: MOLECULAR BEHAVIOR OF A LIQUID, AND THE BROWNIAN MOVEMENT

Instruction for STEP A:

What is the relationship of surface area to the rate of dissolving?

The Dissolving Contest

- Have each child place the same size and variety of life saver or cough drop into his mouth at the same time.
- Have the class make a prediction as to which individual will dissolve the life saver first. Advise them that they cannot chew or break the material into smaller pieces, but must keep it intact as long as possible.
- Noting the time, start the contest.

As different children finish dissolving the life saver, have them raise their hand. Record the time required to completely dissolve the life saver (at least the first, five individuals).

This involvement integrates well with mathematics. The average time it takes to dissolve the life saver, the range, the median, and the mode can be calculated.

Discuss the factors that might have influenced the "winning" of the dissolving contest. Some factors are the size of the mouth, temperature of the mouth, amount of saliva present, acidity of the saliva, pressure applied to the process, the amount of agitation of the life saver while in the mouth, and any affinity the participant may have for the flavor of the life saver.

As you progress through this investigation, it becomes advantageous to make some analogies. Care must be exercised when making these personal comparisons so as not to offend the winner of the dissolving contest: These analogies are:

- mouth compared to a vessel
- saliva compared to the liquid contained in a vessel
- temperature of the mouth compared to the temperature of a liquid in the vessel
- movement of the life saver in the mouth compared to the agitation of the life saver in a vessel due to stirring

Suggested Follow-up Investigations to the Dissolving Contest:

- Compare the rate of dissolving of peppermint life savers to striped or flavored life savers in containers with equal amounts of water. Do they dissolve at the same rate?
- Repeat the dissolving contest, comparing the dissolving rate of a fragmented life saver (four pieces) to a whole life saver.
- Design an experiment to determine the rate of dissolving of three, different, commercially available aspirins. Adjust for differences in their masses.

Involvements extending STEP A:

Fact: a rock in the pulverized state increases its exposed surface area by approximately 10,000 times as compared to its original size.

Will a solid piece of sugar dissolve faster than an equal volume of sugar broken into smaller pieces?

- Using equal volumes of water at the same temperature, drop one, whole cocktail-sugar cube in one container and drop another sugar cube which has been broken into five or six pieces (not pulverized) into another similar container. Compare these two measures to data obtained from dissolving the volume of one pulverized, sugar cube. Which sugar cube dissolved in the shortest period of time? Graph the results. Using a light coating of a soluble glue, glue eight, sugar cubes together to form one large cube. Compare the volume and exposed surface area of this large cube to the volume and total surface area (TSA) of the eight, single cubes. Compare the dissolving rate of the large cube to the eight

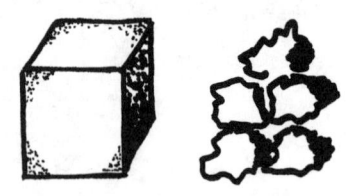

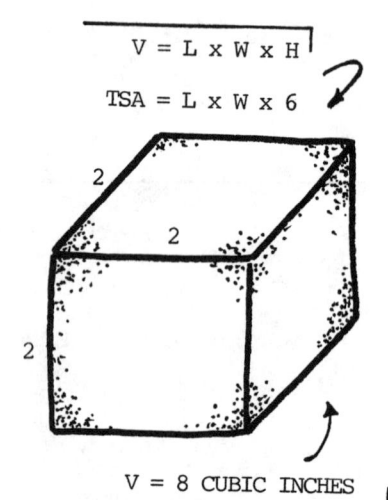

$V = L \times W \times H$
$TSA = L \times W \times 6$

V = 8 CUBIC INCHES
TSA = 24 SQUARE INCHES

V = 8 CUBIC INCHES
TSA = 48 SQUARE INCHES

smaller ones (when simultaneously placed into solution).

. Glue eight, sugar cubes together in any shape you desire excluding a cube. How does the volume and total surface area of this shape compare to the eight single cubes? Dissolve the creative, sugar construct. Compare the rate of dissolving to the original data obtained from the large, glued, sugar cube dissolution.

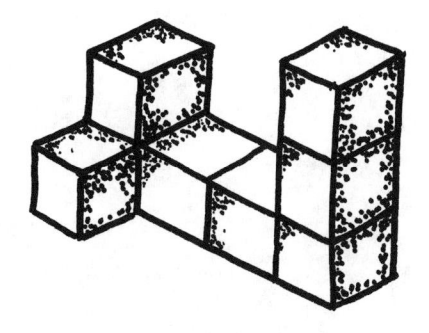

V = ?
TSA = ?

Using the soluble glue can provide interesting results. If the glued cube is allowed to fully dissolve, the glue will dissipate into solution. If you observe carefully, there is a point in the dissolving process wherein the sugar has dissolved but the glue has not yet dissolved. If the glued material is pulled from the liquid before it dissolves, it will be visible as an interesting three-dimensional skeleton. Allow it to dry and it will be quite rigid. If one uses an insoluble glue, the results are more easily preserved. More interesting and elaborate three-dimensional, skeletal arrangements can be generated from other originally created forms.

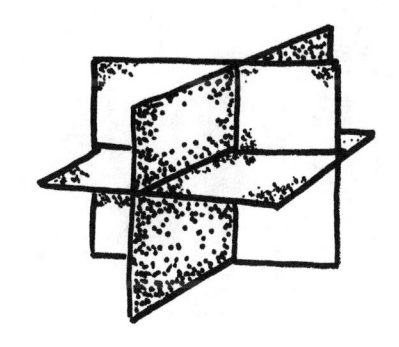

Ice may be substituted for the sugar cubes. However, it is difficult to obtain two, ice cube pieces of exactly the same volume, plus breaking it does not present you with nice even sides for mathematical calculations. Plus, the fact that it melts rapidly present still another problem.

. Select two ice cubes of equal volume. Retain one in the frozen state. Place the other one in a balloon. Let it melt. Squeeze the water down to one section of the balloon. Twist the balloon tightly around the accumulated water until you have fashioned a liquid sphere. Using a pipe cleaner or anything else that will work, fasten the balloon so that the water retains its spherical mass. Hang it pendant inside the freezer compartment of a refrigerator. While the water is still frozen, and after the balloon has warmed up slightly, remove the balloon. The result is a frozen sphere of ice with the same volume as your original, comparison specimen of ice...but the total exposed surface area will be different. How will these two samples compare as to the rate of dissolving? Which will dissolve faster?

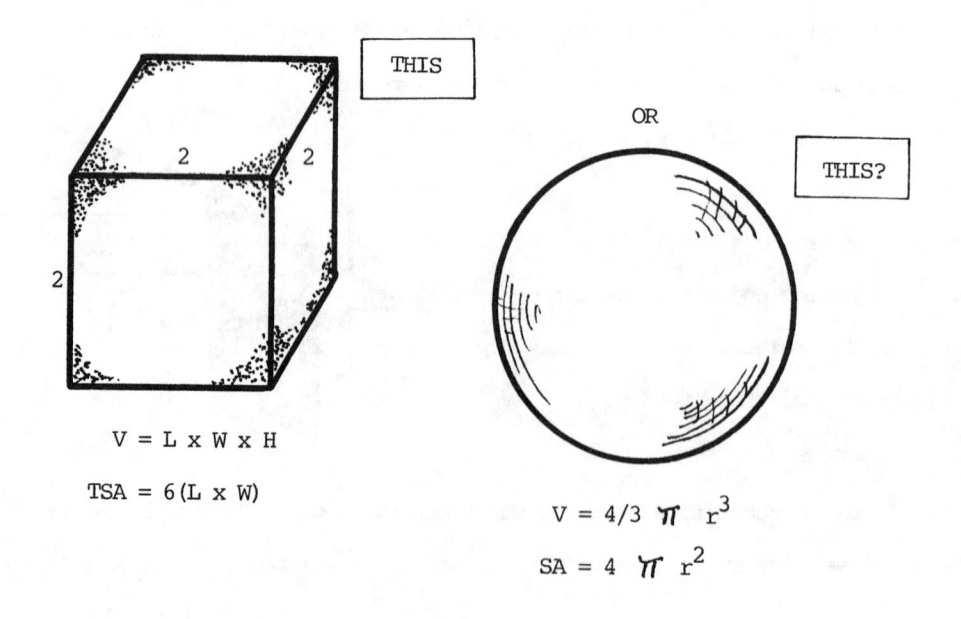

THIS

OR

THIS?

$V = L \times W \times H$

$TSA = 6(L \times W)$

$V = 4/3 \pi r^3$

$SA = 4 \pi r^2$

Instruction for STEP B:

Is the rate of dissolving affected by temperature differences?

- Dissolve a sugar cube in each of three containers filled with equal volumes of water. Vary the temperature of the water in each container - room temperature, cold, and hot. Record the data. Graph the results.

Is the rate of dissolving affected by liquid differences?

- Dissolve a sugar cube in each of three similar containers filled with equal volumes of liquids. Vary the type of liquid - tap water, rain water, and salt water. Record the data. Graph the results.

Does stirring affect the rate of dissolving?

- Dissolve a sugar cube in each of two containers containing equal amounts of water at similar temperatures. Do not disturb one container. Stir the contents of the other container without striking the sugar cube. Observe. Record the results. Conclude.

Involvements extending STEP B:

Is the rate of dissolving affected by materials included in the formation of ice cubes?

- Prepare ice cubes made with fruit juices or sugared water. Compare the dissolving of these to plain, frozen-water cubes of the same volume and the same total, exposed surface area.

Does the amount of liquid in which a soluble material is placed, affect the rate of dissolving?

- Assemble two containers - one with 100 ml of water and one with 500 ml of water. Dissolve one sugar cube in each. Record the time it takes to dissolve each cube. Conclude.

Suggested Follow-up Investigations to STEP B:

Design an experiment to establish if the rate of dissolving is affected by sound waves passing through solution.

If equal amounts of water are used, would a soluble material dissolve more quickly if positioned at different depths?

Involvements extending STEP B:

Will a two inch cube of ice melt faster than two, cubic inches of ice?

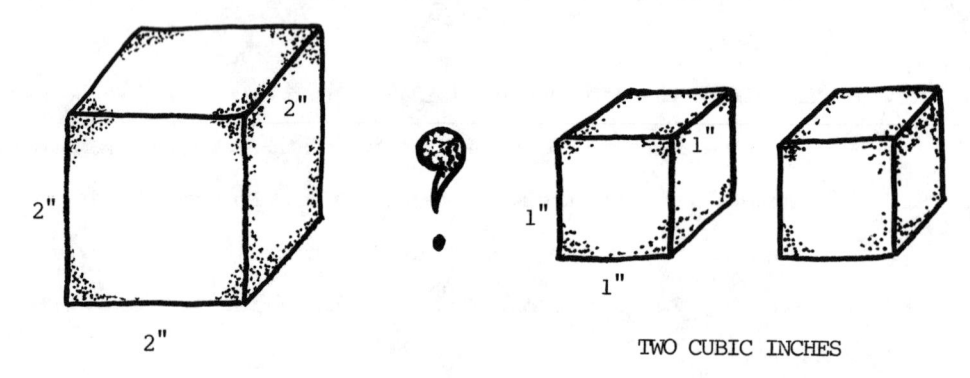

TWO CUBIC INCHES

Which is superior for cooling drinks - crushed ice or an equal volume of a single chunk of ice?

If you double the linear dimensions of an ice cube, what do you do to its volume?

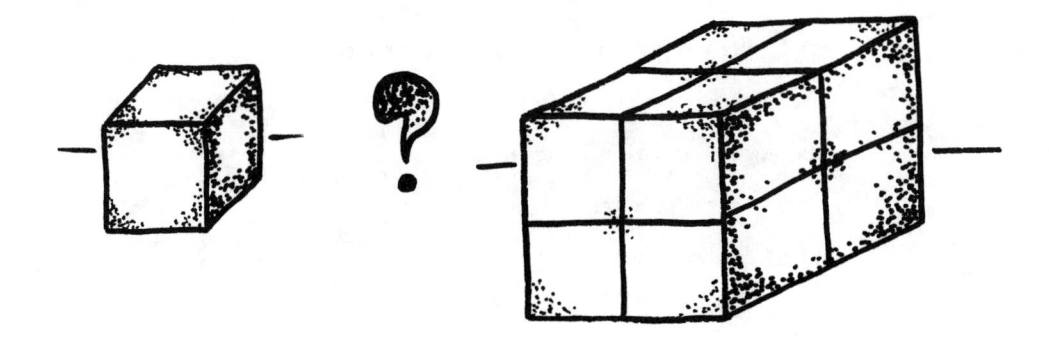

Design an ice cube that will melt rapidly.

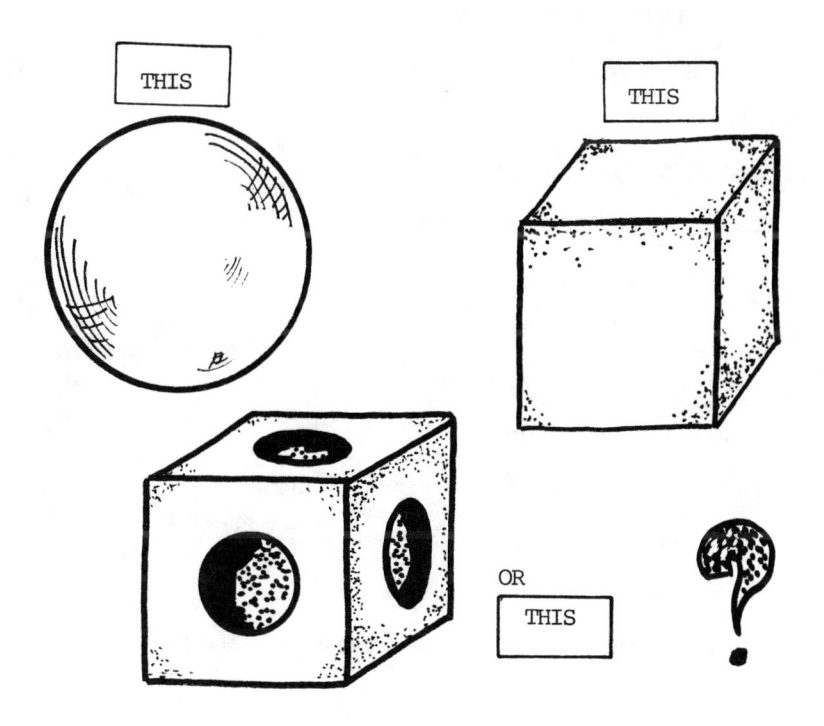

THIS

THIS

OR THIS

169

Instruction for STEP C:

Are phase changes reversible?

. Place an ice cube in a container. Let the ice cube melt. Let the water formed by the melted ice cube evaporate. Place an empty container in the freezer compartment of a refrigerator. Let it become extremely cold. When this has taken place, take the empty container out and place it on one pan of an equal arm balance. Adjust the other arm so that the scale is in balance. If the container is cold enough, and if the humidity is right, the container should fog and begin to sweat. The balance should reflect this added mass and shift out of balance. The amount of weights that it takes to bring the scale back into balance should be equated by an equivalent mass of water. Pour this equivalent amount of water into a plastic bag and freeze it. Thus the cycle has been completed. We have gone from a solid to a liquid, to a gas and from a gas back to a liquid, and then back to a solid. Are all phase changes reversible?

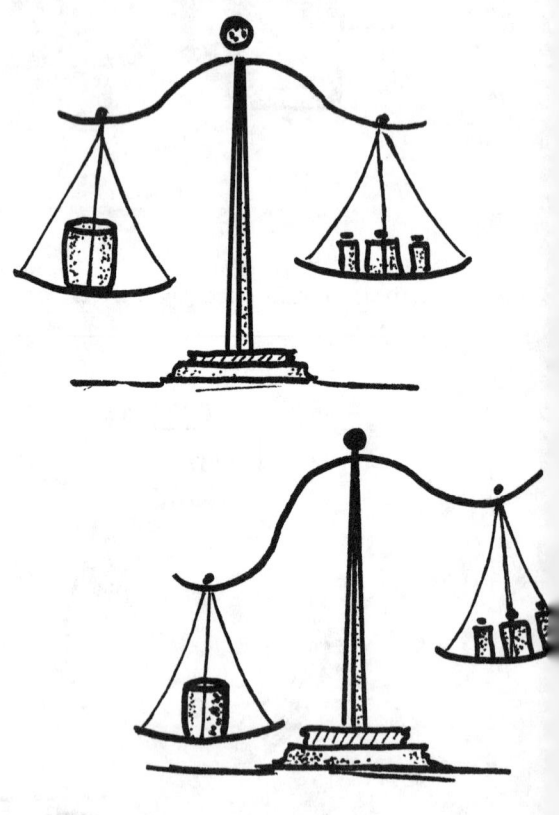

Design an ice cube that will melt slowly.

How does volume and total surface area apply to the cooling of the human body on a hot summer day?

Suggested Follow-up Investigation to STEP C:

- Pharmaceutical businesses manufacture medicinal capsules that contain three or more medicines; and they want these medicines to work at different intervals in different portions of the body. Sometimes these are referred to as "spansules" because they dispense medicine throughout the body over a span of time. From your knowledge of area, volume, and the rate of dissolving, design a capsule and its components so that the spansule phenomena could occur.

- A nice extension to phase changes would be a consideration of what happens to mass and volume as one goes from a gas to a liquid to a solid and vice versa. This seems to lead into an ancillary study of crystal formations as one considers rapid freezing vs slow freezing of a liquid to a solid. Add salt or sugar to hot water until the water is saturated. Place half the solution in a warm, dry environment. Place the remaining half in the freezer compartment of a refrigerator. Compare the two final crystal formations.

Fact: For a given volume of a substance, the greater the surface area exposed to a possible phase change the more rapid will be the phase change.

Instruction for STEP D:

Do camphor crystals placed in a container change to a liquid and then to a gas?

- Place camphor balls in an open container. Observe and record changes in mass of a period of time. Record your results.

Instruction for STEP E:

How much of any one substance can a fixed volume of water hold?

- Place a known amount of tap water in a container. Add salt to it. Keep stirring and adding salt until no more salt will go into the solution. The salt is seen to precipitate or settle out. Determine the mass of the salt before you start and the remaining amount retained when the precipitate is first noticed. The difference is the amount of salt that is held in solution.

How does the amount of salt that is held in solution vary based on temperature differences?

- Place a known amount of tap water in three different containers.

Each container will have a different water temperature - room temperature, cold, and hot. To each container while stirring, add salt until no more will go into solution; and the salt is observed precipitating out appearing on the bottom of the container. In each case, determine the mass of the salt before you begin and again at the first sign of a precipitate. The differences between these two amounts of salt is that amount of salt which in each case has passed into solution.

What happens to any observable salt precipitate when warm or hot water cools?

. Prepare two super-saturated solutions of salt, one in lukewarm water and one in very hot water. Allow these two containers to cool. Observe the precipitate in each instance. How do these amounts compare?

Fact: Heating water will expand the space between water molecules. In this condition, the water is capable of containing more salt than it could in cold water where the water molecules are packed closely together. A solution is termed super-saturated when it contains more salt than it would normally contain under room temperature. Water, at room temperature, containing the maximum quantity of salt that it is capable of containing prior to any salt precipitating out, is termed a saturated solution. Hot solutions in a super-saturated state precipitate out salt as the water cools. When the water cools, the molecules of water contract and are not able to hold as much salt in solution. Excess salt precipitates out.

An Associated Demonstration

Fill a beaker with marbles to a predetermined level. The marbles represent water molecules. Spaces exist between the marbles. Pour BBs (to simulate salt) into the beaker. The BBs fill the voids between the marbles When the voids are filled the solution is said to be saturated. When no more BBs can fit into the spaces, the BBs accumulate at the top of the beaker. With no more room available, the BBs in a sense precipitate out. In water, excess salt, which has filled all the voids between the water molecules, accumulates on the bottom of the beaker.

If one could envision the marbles moving further apart as a result of being heated, one could envision that more BBs (salt) could fill in the enlarging voids between the marbles.

Is evaporation related to exposed surface area?

Use two vessels, cylindrical or otherwise, providing of course, that the two vessels have similar openings and consistent shapes. Calculate the surface area exposed to the air at the opening of each vessel. Compare these. Place equal amounts of water in each vessel, place them in similar environments and have the students predict the rate of evaporation of water in one vessel compared to that of the other.

This activity can be extended by comparing the amount of evaporation of equal amounts of water in two vessels both of which have the same opening; one whose shape is consistent, compared to one whose shape is inconsistent. A cola bottle with its hour-glass shape would be a good example of an inconsistent shape. Does a constantly changing water-surface area alter the evaporation rate? Compare the two rates.

Can evaporation be measured by weighing?

Weigh empty containers. Add equal amounts of water to each container. Weigh them. Place these containers in different environments. After a time interval weigh the remaining water and vessel. This weight subtracted from the original, total weight will give the amount of water evaporated. The rate of evaporation may be determined knowing the time and the quantity of water evaporated.

This involvement could be extended by using a canvas bag filled with water suspended outdoors in the sun on different days of highly contrasting humidities. This would give us a measure of evaporation as related to humidity. By weighing the sample before and after the experiment and permitting equal time exposures on both days, a comparison can be made and conclusions drawn as to the rate of evaporation as influenced by humidity. Humidity data are usually published in local papers.

An Associated Demonstration: Brownian Movement

Molecular motion is difficult to observe directly, however, its effects can be observed without duplicating Robert Brown's historic pollen-grain demonstration. Any fresh, carbonated drink wherein the escaping carbonated, gas bubbles are visible, can be used to simulate Brownian Movement. Most antacids will also provide a fair representation of the effect of molecular motion as particles of the liquid escape into the atmosphere.

As a direct outcome of the discussion and involvement with saturated, supersaturated solutions, and molecular voids, the topic of porosity and permeability evolves naturally.

> Fact: Porosity in a rock is commonly reported in percent, showing the volume of pores compared to the volume of the rock plus pores. Permeability is defined as the volume of a fluid that can pass through a given volume of material under some prescribed unit of pressure in a given unit of time.
> Some earth materials are very porous but not necessarily permeable. These earth materials can hold a great deal of water, but they do not allow water to pass through them.

How can the permeability of a sample of soil be determined?

- Use four, one-half pint milk cartons of equal shape. Fill one carton three-fourths full with washed and dried pea-size gravel. Fill the second carton to the same level with a washed and dried fifty-fifty mixture of sand and pea-size gravel. Fill the third carton to the same level with all washed and dried sand. And, fill the fourth carton to the same level with washed sand, only

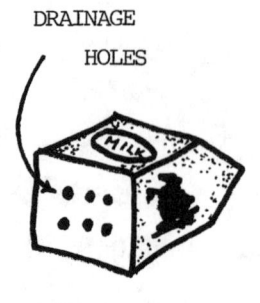

DRAINAGE HOLES

PEA SIZE GRAVEL

50/50 SAND/PEA GRAVEL

sandwich in the middle of the sand a one-quarter inch layer of clay. Pack the clay tightly and press it so that it is sealed against the sides of the container. Using a nail, punch a hole in the bottom of each carton. Place the four cartons over a catch basin or some other waterproof container to catch runoff water. Slowly add sufficient water to each carton until you see water dripping out of the bottom of the carton. Record the time this takes to occur. Also record the differences between the amounts of water put into each system and how much ran out. Collect and graph the data.

Provocative Question: Is melting the same as dissolving?

A Model for Spatial Development

(SIMPLE TO COMPLEX)

- Pattern Blocks
- Geoblocks
- Attributes and Games
- Spatial Art
- Spátial Processes
- Block Letter Cut Outs
- Tangrams
- Match Stick Problems
- Nine Dot Problem
- Circle and Arrow Orientation

    (rotation about two axes)
- Triangle (rotation and transformation)
- Folded, Punched Paper
- Embedded Figures
- Tessellations
- Shapes and Things
- Cubes with Six Different Face Patterns
- Three Dimensional Plotting

    plus Polar Coordinates
- Three Dimensional Projection
- Six Crystal Systems (axes of symmetry)
- Three Dimensional Cross-Sections

    (faulting, folding, and subsequent erosion)
- Three Dimensional Subsurface Geologic Models
- Rubic Cube

SPATIAL PROBLEMS TO ACCOMPANY SCIENCE INSTRUCTION

Elementary school children are disadvantaged when it comes to experience in three-dimensional considerations in science. Instruction in science in the elementary school is often solely confined to paper and pencil desk work (planar) by the children and blackboard (planar) work by the teacher. For many children, the world is a constantly reinforced two-dimensional world.

Improvement in spatial manipulations and interpretations, is contingent upon a sustained personal involvement by teachers and children in spatial problems. Numerous elementary school science programs provide such opportunities. Experience with these manipulations, some relatively simple, some complex, should be provided for those children who lack a feel for time/space and motion events. Some spatial elementary science considerations are:

. Pattern Blocks

Teacher's Guide for Pattern Blocks, 1970[*] Elementary Science Study, Webster Division, McGraw-Hill Book Co., N.Y., N.Y.

Pattern Blocks are blocks made of wood about 1/8 thick and one inch long (except for the two inch side of the trapezoid). There are 250 blocks, each having a characteristic shape. Each characteristic shape is similar in color: 50 green triangles; 50 red trapezoids; 50 wide blue diamonds; 25 orange squares; 25 yellow hexagons; and 50 narrow diamonds (natural color). Three mirrors are also included with these blocks. These blocks have many uses. They can be used to establish patterns, symmetry, mirror symmetry, symmetry or rotation, perimeter and area, arrangement and rearrangement, angle measurement, series and sequences, and much more.

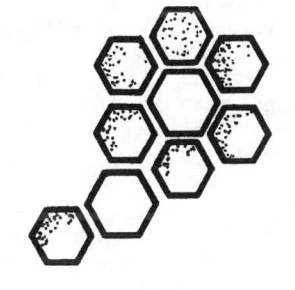

- Geoblocks

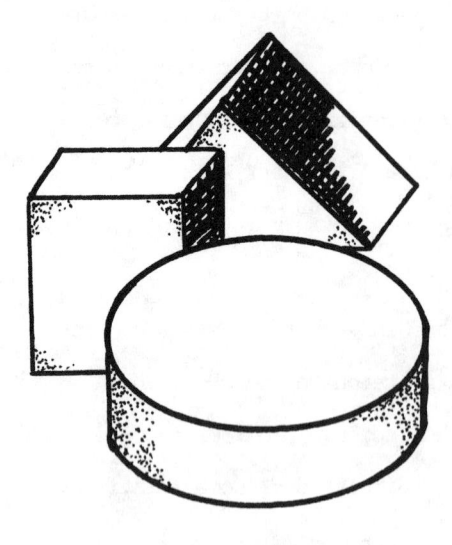

Teacher's Guide for Geo Blocks, 1969[*]

Geo blocks include 330 pieces of unfinished hardwood cut into a wide variety of shapes and sizes. The smallest is a half-inch cube; the largest is a 2" x 2" x 4" oblong. The particular shapes and sizes were selected for their volumetric relationships and for their usefulness in making intricate constructions. As a result, there are only three blocks in the set which cannot be made by putting together other blocks.

Problem cards are provided. Some cards show all the faces of a particular block drawn in outline. Others are photo sets of different views of blocks, for example, side, top, etc. Still other cards show several views of a constructed block structure drawn in outline. In each case the students duplicate the structure or design by manipulating the blocks.

These activities can be extended to include experiences with orthographic projections, selection of views, symmetry and asymmetry.

- Attributes and Games

Teacher's Guide for Attribute Games and Problems, 1968[*]

While many segments of Attribute Games and Problems are helpful in providing experiences in spatial relationships, "A" blocks (32 wooden blocks, all different) are particularly useful. The concept of color mapping, size mapping, and shape mapping is introduced. This is extended to color and size mapping, shape correspondence, shape and color mapping, mirror image, and size mapping with rotation.

Color Cubes (60 cubes in 6 colors) another segment of Attributes Games and Problems are also applicable for promoting experiences with rigid transformation. Colored cubes are arranged in a set pattern. Segments of the pattern are rotated. Students are challenged to reconstruct the steps leading to the location changes.

- Spatial Art

Creative Publications, P.O. Box 10328, Palo Alto, CA 94303

String Sculpture by John Winter

> String sculptures are unique designs constructed entirely with straight lines. Simple or highly complex constructs can be assembled. Simple geometric figures are often used as the basic outline. Equally spaced points are marked off along each edge. String is stretched between points in a well defined order. The envelope (outermost edge or surface) of the straight line segments forms a surface which seems to be curved. The resultant envelope can be altered by changing the "rules" by which points are connected. These curves can be two dimensional or three dimensional depending on the design which is employed. Three dimensional string sculptures using intersecting plane surfaces can also be constructed.

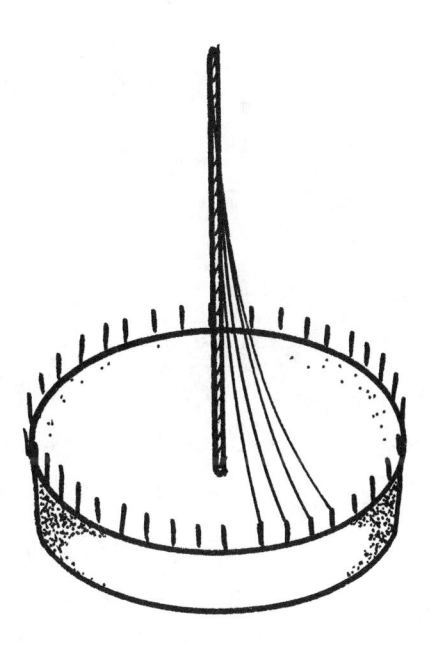

- Spatial Processes

Science-A Process Approach, Commentary for Teachers, Ginn & Co.,

Boston, MA 1970

> Using space/time relationships is the process that develops skills in the description of spatial relationships and their changes with time. It includes a study of shapes, symmetry, motion and rate of change. Seventeen exercises in this process are included in the Science -A Process Approach materials. These can be grouped into four categories. Shapes, Time, Direction and Spatial Arrangement, and Motion and Speed.
> Exercises are provided in the generation of three-dimensional shapes by rotating, in space, two dimensionally shaped objects. Projective transformations, wherein the identification of three-dimensional shapes from the projections of their two-dimensional shadows, are also included. Under the process skills of "interpreting data," Science-A Process Approach provides further spatial exercises. These activities are initiated through an introduction to three dimensional graphing.
> Involvement in these basic activities will not guarantee the de-development of perceptive observers in science, but it should broaden the base for the comprehension of spatial phenomena.

. Block Letters

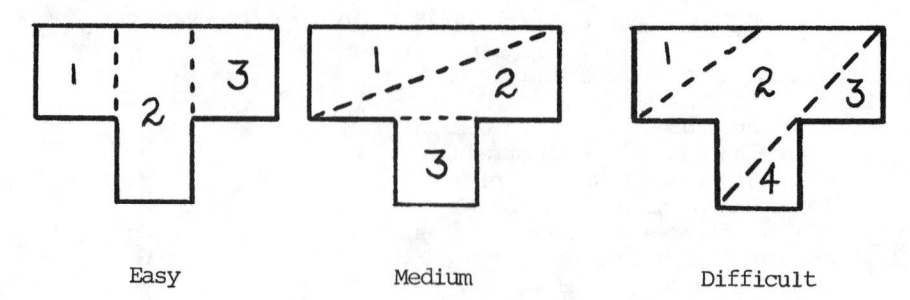

                Easy   Medium   Difficult

Cut the block "Ts" along the dotted lines. Shuffle the pieces of any one set. Reassemble the pieces to form the letter "T." This activity can be extended to almost any block-formed letter of the alphabet. Some letters, by their configurations, are more suitable for this exercise than other letters.

This activity can be extended to other geometric shapes, for example, "How should the cross be cut so that the pieces can be assembled into a square?" *

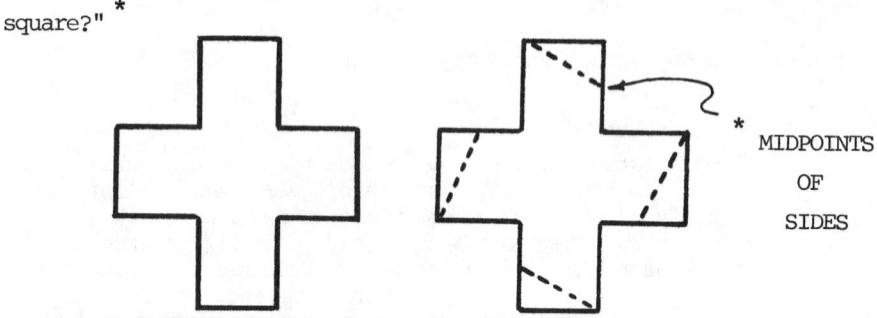

* MIDPOINTS OF SIDES

. Tangrams

Teacher's Guide for Tangrams, 1968*

> Tangrams is a geometric puzzle, it consists of a square divided into seven geometric shapes: two large triangles, a medium triangle, two small triangles, a square, and a rhomboid. Using these pieces a great number of geometric arrangements can be made. Early instructional manipulations are confined to fewer pieces and simple printed designs are furnished mapping the geometric pattern. Later, more intricate patterns are introduced and the full seven pieces are utilized. Multiple solutions are achieved for many of the problems. Students move from fingertip solutions, to visual ones, to purely analytic ones.

Tangrams

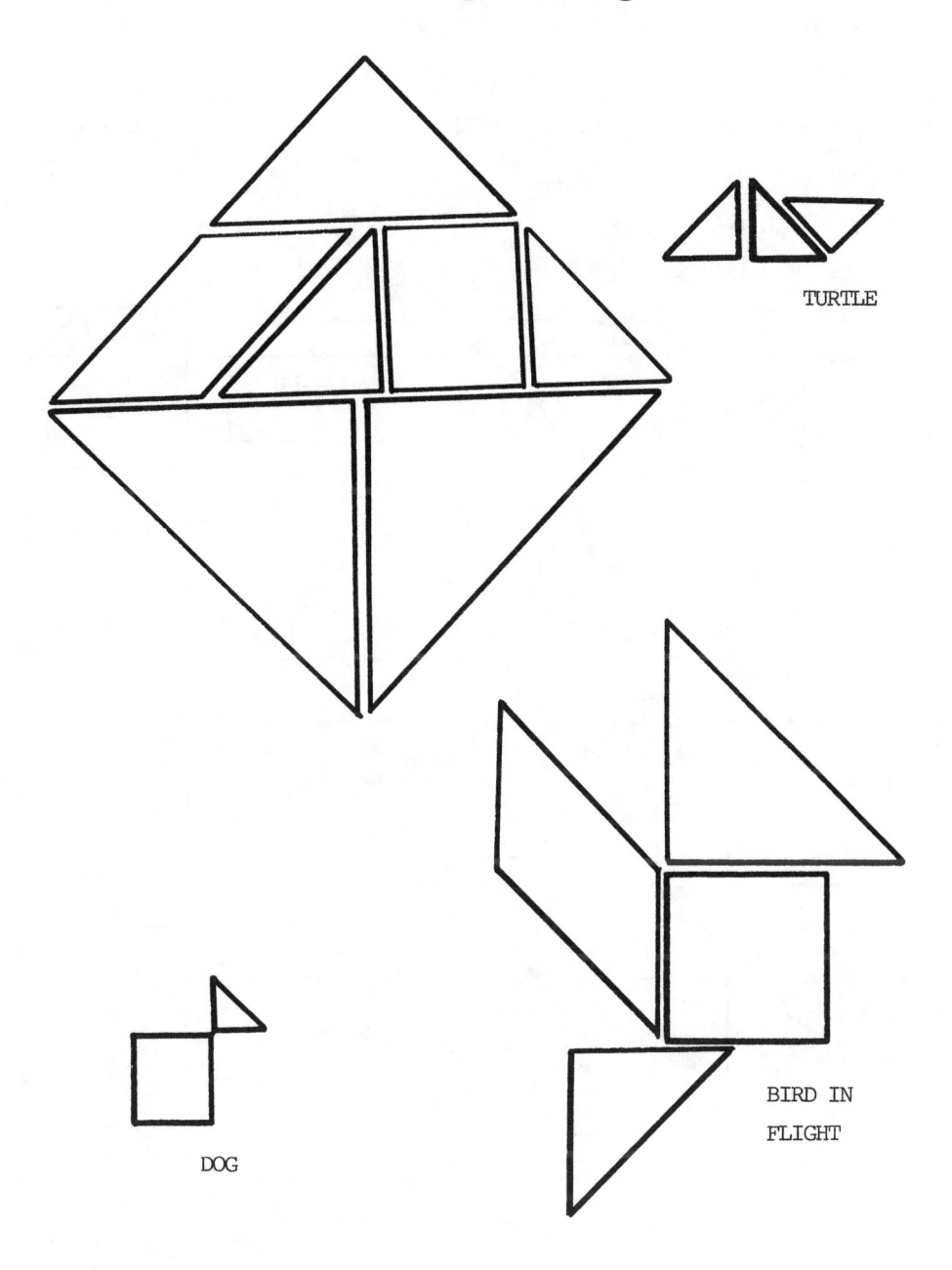

TURTLE

DOG

BIRD IN FLIGHT

183

. Variations on the Tangram Theme

Each square is made up of the pieces to the right of it. Either make cut-outs to fill in the area of the square or draw appropriate lines in the square to show how the pieces nest within the square completely filling the area

A - E...easy to difficult

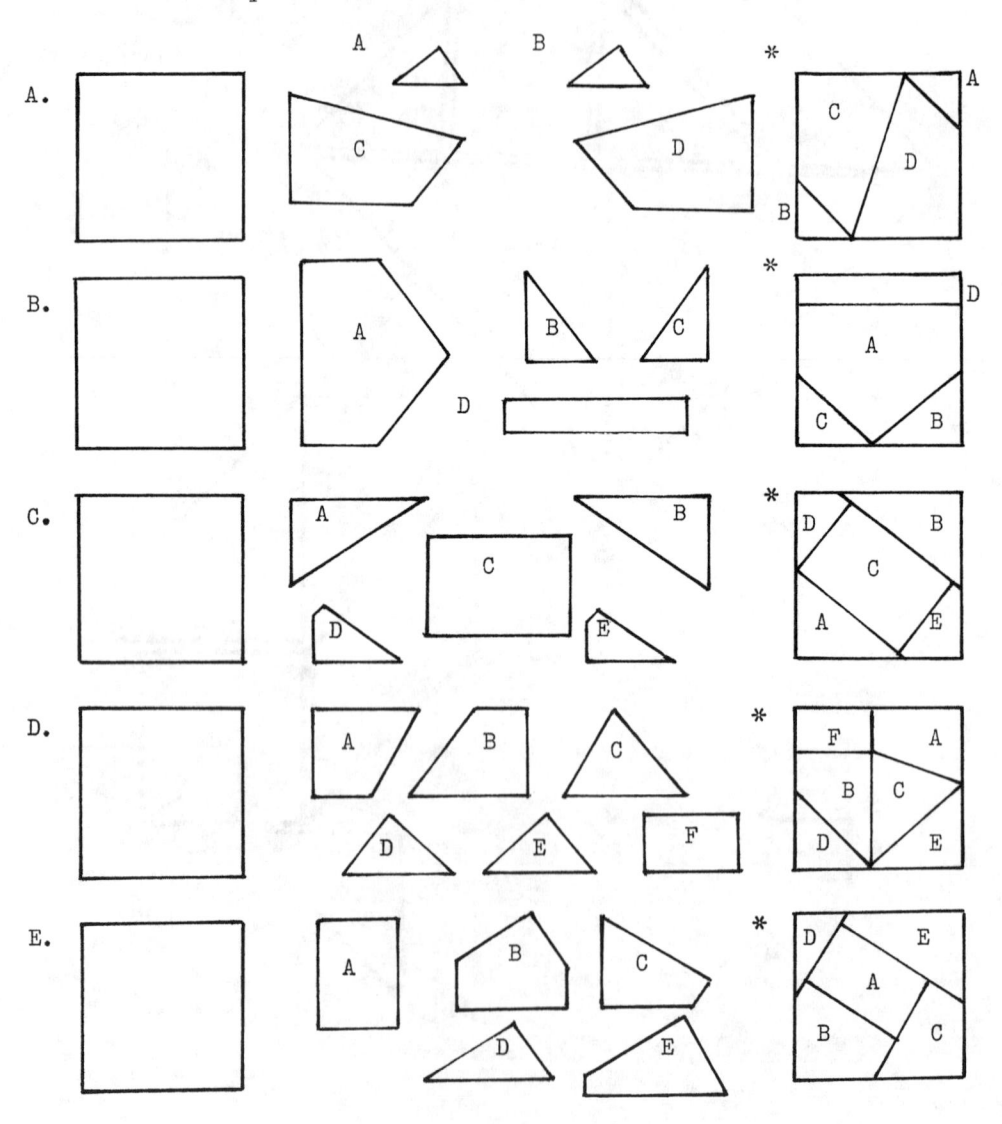

- Match Stick Problems

Make six squares with seventeen toothpicks. By removing five toothpicks, three contiguous squares should remain.

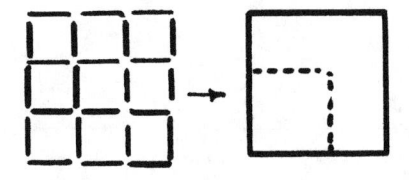

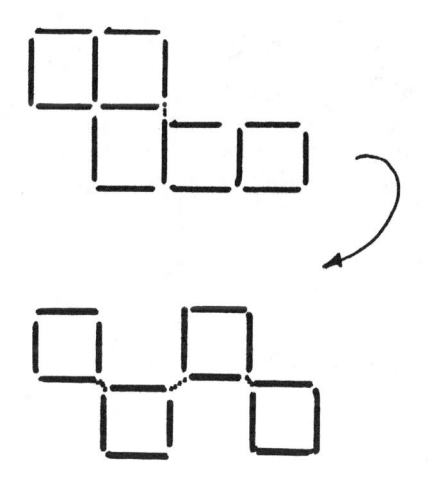

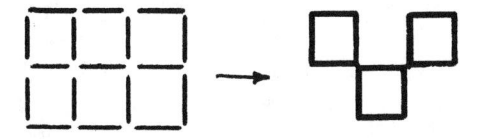

Using twenty-four toothpicks, form nine squares. Now remove eight toothpicks so that only two squares remain.

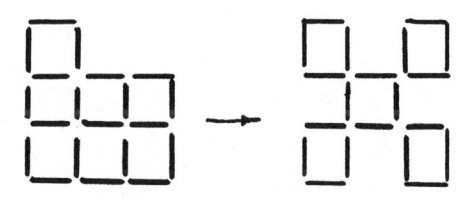

Arrange twenty toothpicks to form seven squares. Remove three toothpicks and replace them so as to form a construct of only five squares -- all the same size, and touching each other at some point.

Move two and only two toothpicks and re-incorporate them back into the system (don't place these alongside or on top of another toothpick), reducing the number of squares from five to four squares. Any one of the four remaining squares must be the same size as any one of the original five squares. Also the four squares must be contiguous.

- Remove eight toothpicks so that three squares, not necessarily all the same size, remain.

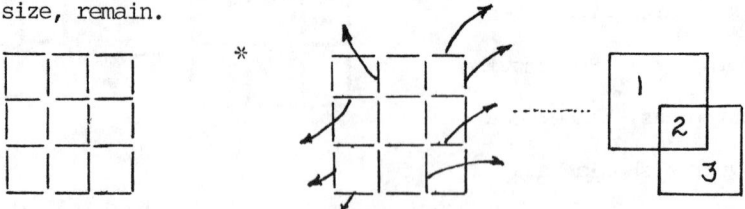

- Move two toothpicks and have eleven squares remaining.

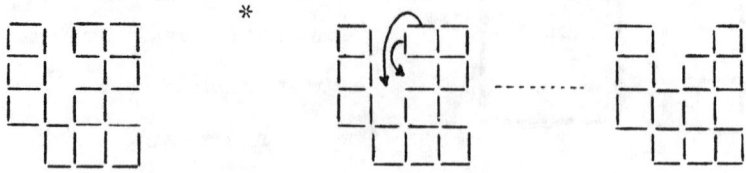

- Move two toothpicks so that you will have eight squares of the original size.

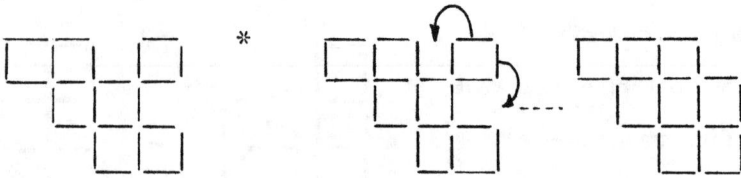

- Remove four toothpicks and have only five of the original squares remaining.

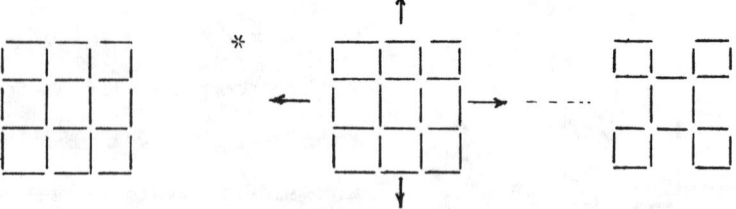

- Move two toothpicks in such a manner that you will have six squares of the original size remaining.

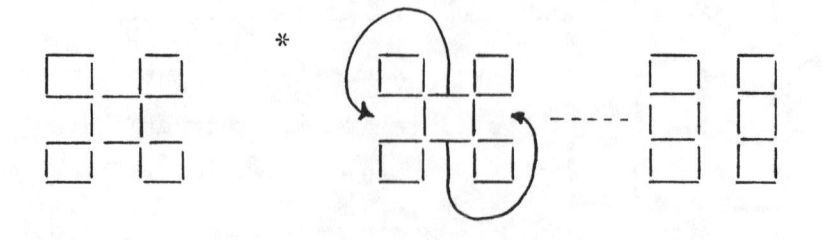

. What is the minimum number of lines needed to separate the sheep into individual compartments?

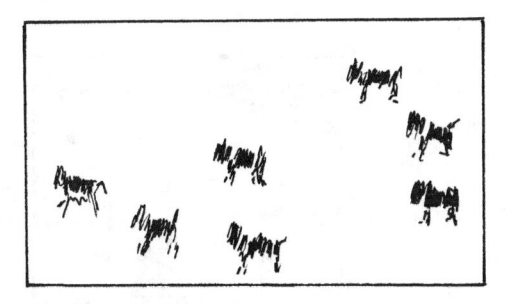

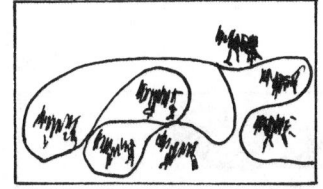

* One if it curves and crosses itself.

. Surrounding the large gazebo are four large houses and four little houses. The people living in the little houses want to build a fence to keep the people in the large houses away from the gazebo. But the people in the little houses want to be able to get to the gazebo themselves. Draw a line to show where a fence could be located to accomplish this.

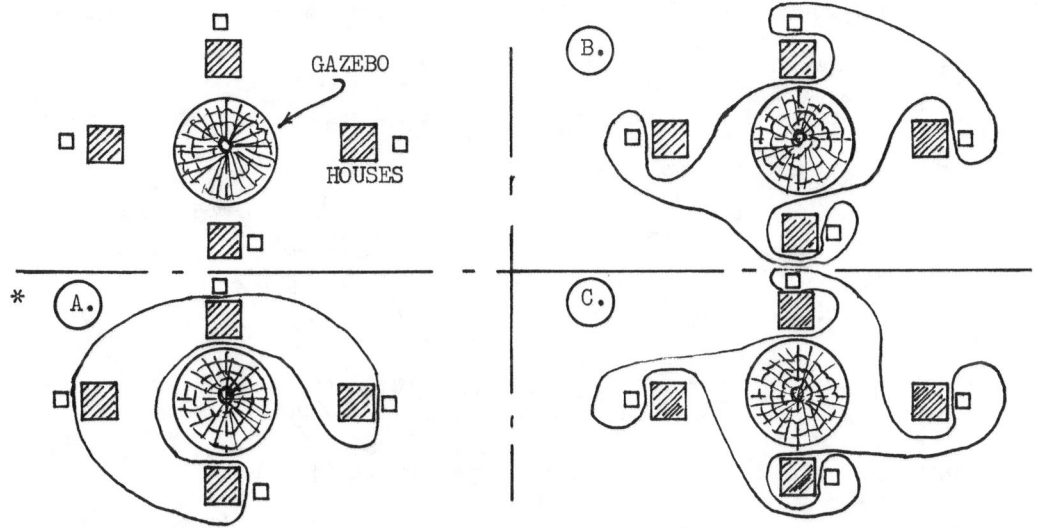

* A.

B.

C.

. Mr. Freddy Friendly has a square piece of ground. In one quarter of this land he built a four family apartment house. Which was rented to four different families. Being friendly, he told them that if they could divide the remaining three fourths of the property into four equal plots, alike in shape, and each containing one of the four apple trees he has planted, they could each have their own backyard with no increase in price. How could they accomplish this?

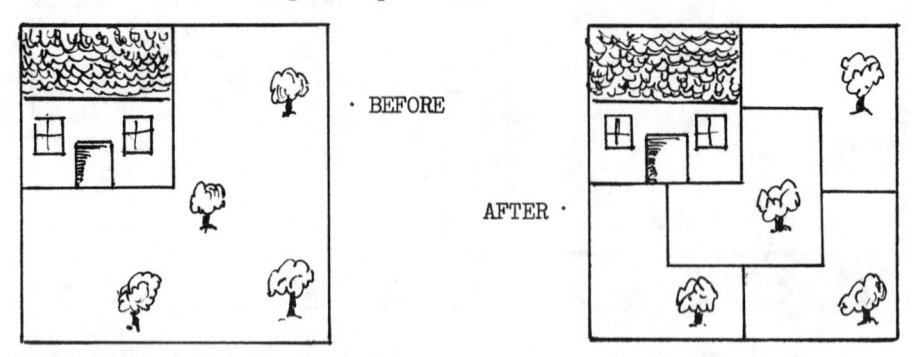

BEFORE

AFTER

- The action will tell you.

  Arrange the materials as shown in the sketch. What kind of a path will the pencil trace if the pencil is held upright and steady while the paper is pulled to the right? What will the path look like if the pencil is moved down the paper as the paper is moved from left to right? What would the curve of the pencil trace make if the pencil was moved in the same manner, but the paper was moved from right to left?

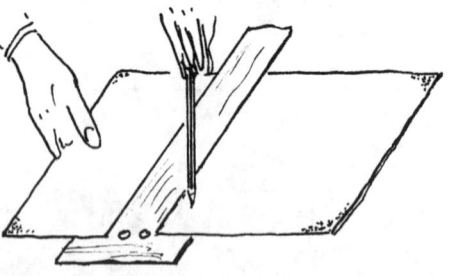

- Here is a top and front view of the same object. Both views are drawn to the same scale. Make a sketch of what a three-dimensional view would look like. (There is more than one answer to this problem).

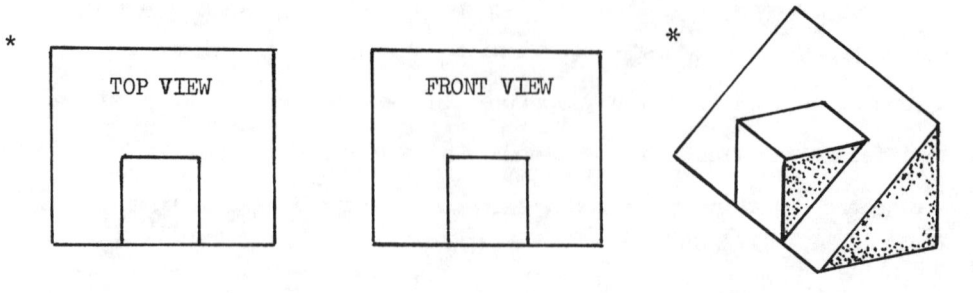

  * TOP VIEW   FRONT VIEW   *

. Ten coins are arranged in a pyramid order. Reverse the order of the pyramid by moving only three coins.

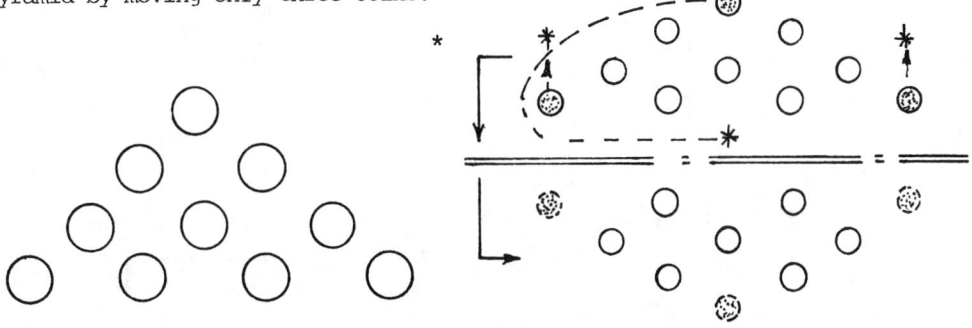

. Each of these two geometric shapes has been divided into three equal parts of the same shape. Divide each of these geometric shapes into four equal parts of the same shape.

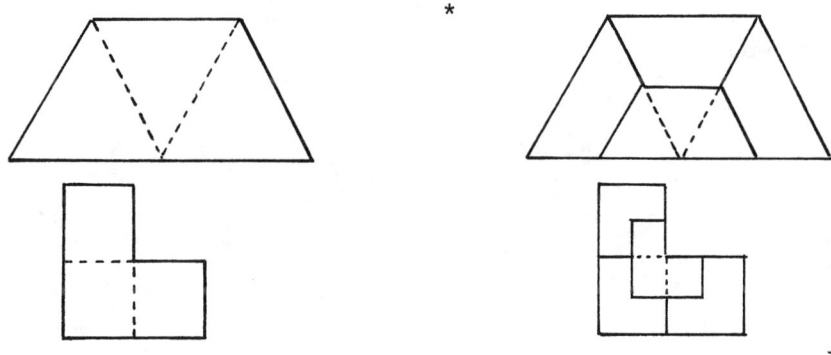

. Which is it a cube with a piece missing or a box in a corner?

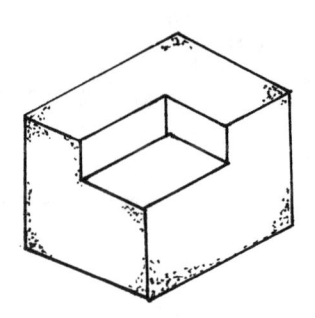

* Either one!

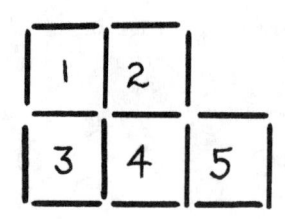

Two farmers purchased a five acre parcel of land. They wished to lay out a straight fence that would divide their purchase into two equal plots. How might they have accomplished this?

*Imagine the square on the right being divided equally by a verticle line. Project this line and think of one-half of the cube being placed on top of the other half. The oblique line will then divide the area into two equal parcels of land.

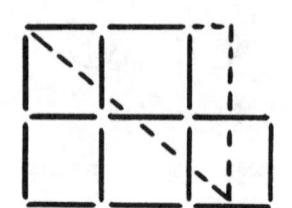

. Nine Dot Problems

Nine dots are arranged to form a square. Draw four straight lines so as to cross every dot. Do not cross any dot more than once, nor retrace any line, nor lift the pencil from the paper until all nine dots have been crossed.

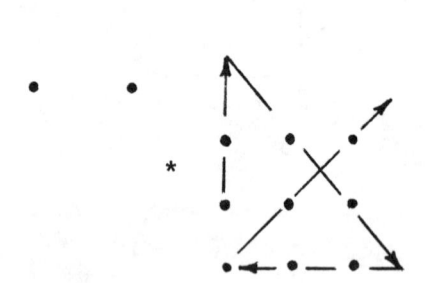

. Circle and Arrow Orientation (rotation about two axes)

From a piece of cardboard, cut out a disc about two inches in diameter. On one side of the disc draw an arrow. On the reverse side draw a similar arrow, but position it ninety degrees out of phase (one arrow at twelve o'clock and the other at either nine or three o'clock). Rotate the disc between your thumb and forefinger in an east-west direction (around a vertical axis) asking an observer, who only sees one side of the disc, where the arrow is on the reverse side which will be the side facing you. Continue to alter the position of the arrow by rotating the disc in clockwise or counter-clockwise positions as well as shifting it in an east-west or west-east direction. As you move through various manipulations, challenge the observer to again state the location of the arrow, on the reverse side (as viewed by you) as a consequence of various disc movements. Ask, "Where is it now?" Flip, rotate, or rotate and flip the disc. At every pause, continuously asking "Where is it now?"

To add a new dimension to the activity, change the direction of the rotation. Rotate the disc in a north-south or south-north direction (around the horizontal axis). Again, challenge an observer to describe where the hidden arrow is as you continue to vary the clockwise and counter-clockwise rotations along with horizontal rotations. Start simple, then move to more complex maneuvers. Mix clockwise and counter-clockwise rotations alternately with horizontal and vertical rotations.

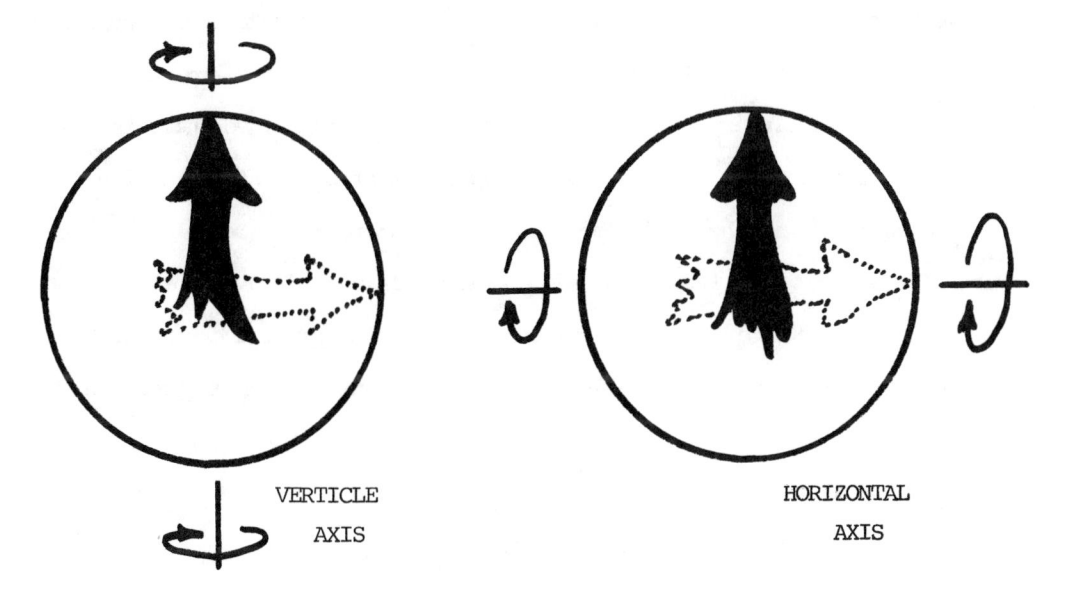

VERTICLE AXIS

HORIZONTAL AXIS

- Triangle (rotation and transformation)

    From a Manila folder, cut a triangle. Label the angles, A, B, and C. Flip it over. Where is angle "A" now? Where is angle "B?" Where is angle "C?" Select any angle as a pivotal angle. Swing the triangle through a selected arc (ninety degrees). Where are the various angles located now? At the conclusion of successive flips and various rotations, challenge an observer to describe the position of one angle, several angles, or the positional relationship of one angle to another.

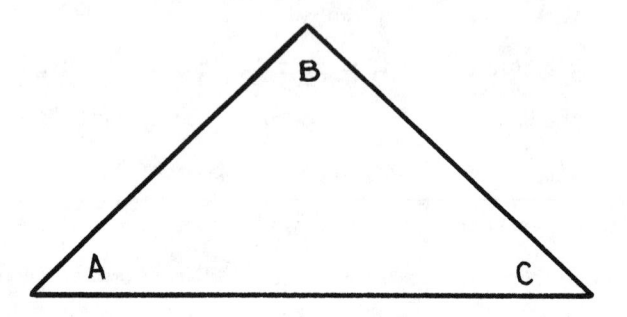

- Folded, Punched Paper

    Cut from paper, five rectangles approximately 5½" x 8½". Fold one sheet in half. Using a hole puncher, punch a hole in the center of the folded sheet. Infer where the holes will appear when the folded sheet is opened to its original size. Take another sheet of paper, fold it twice. Punch a hole in the middle of this. Infer where the holes will appear on the paper when this double-folded sheet is opened to its original size. Repeat this for each remaining sheet of paper increasing the number of folds one more time as you progress to new sheets. Infer the results in each case. Can you predict the results for a sixth sheet without actually punching it? Can you evolve a formula relating the number of folds to the number of punches?

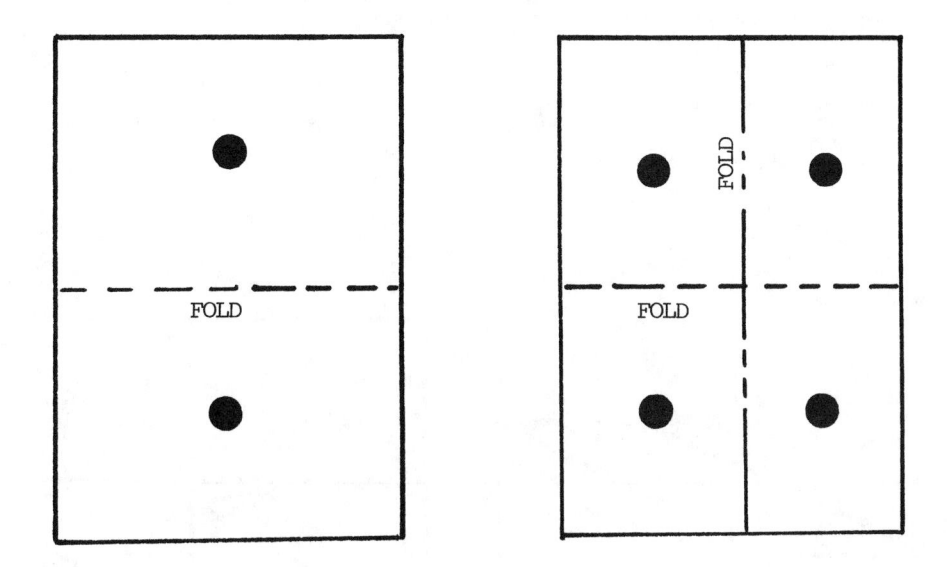

This involvement can be escalated by varying the nature of the fold or the timing of the punched holes in relation to the sequence of folds. The folds and the punches can be established in any sequence to challenge the participants in predicting the final patterns.

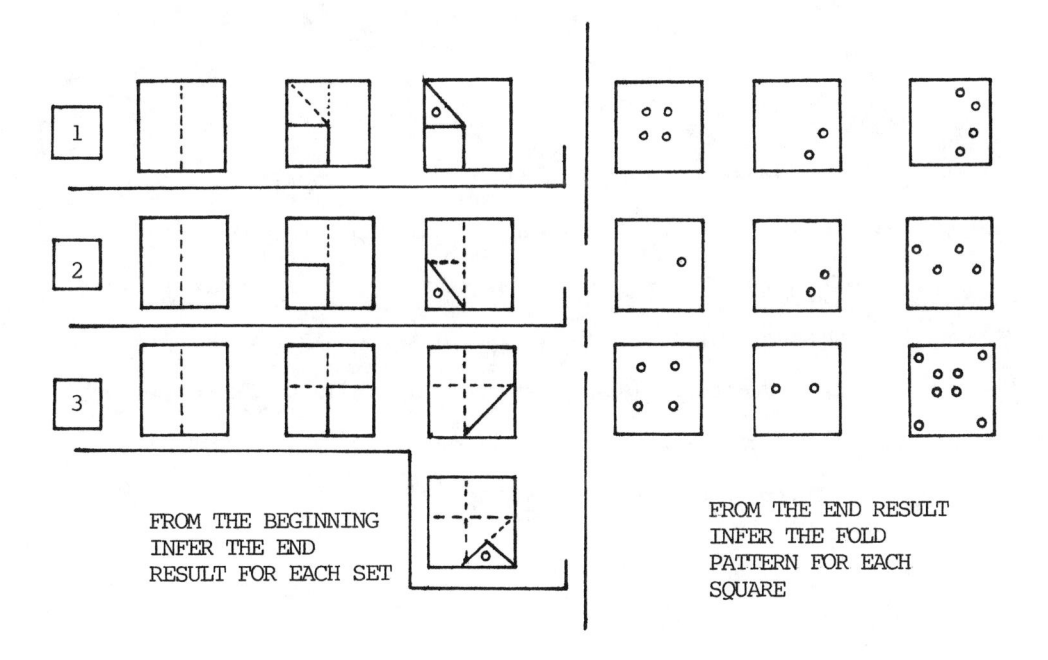

FROM THE BEGINNING INFER THE END RESULT FOR EACH SET

FROM THE END RESULT INFER THE FOLD PATTERN FOR EACH SQUARE

- Embedded Figures

    Find the simple shape in the aggregate shape.

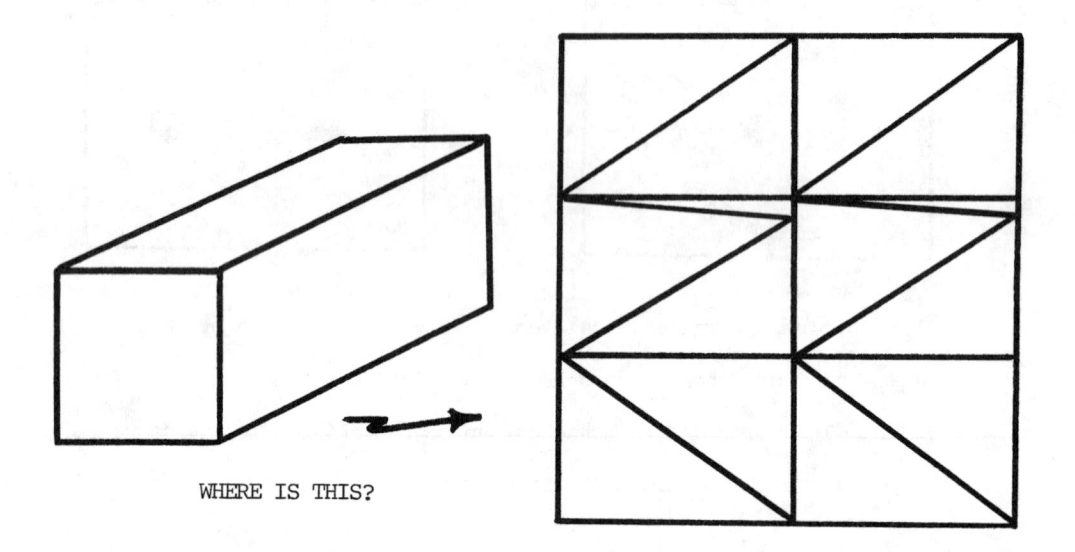

    WHERE IS THIS?

- Tessellations

    Tessellations, sometimes called a tile setter's nightmare, are easy and pleasant to complete. They require little equipment but generate a great deal of spatial experience. Layouts or pattern-trace sheets can be arranged in verticle and horizontal rows or in oblique rows. Each variety accomodates distinct shapes. The child traces or places cut-out shapes within the confines of a set of points. Adjoining drawings or pieces are nested to generate various patterns.

Oblique Line Plotting

Will any one piece nest to form a continuing pattern? Will one or more pieces nest to form a continuing pattern?

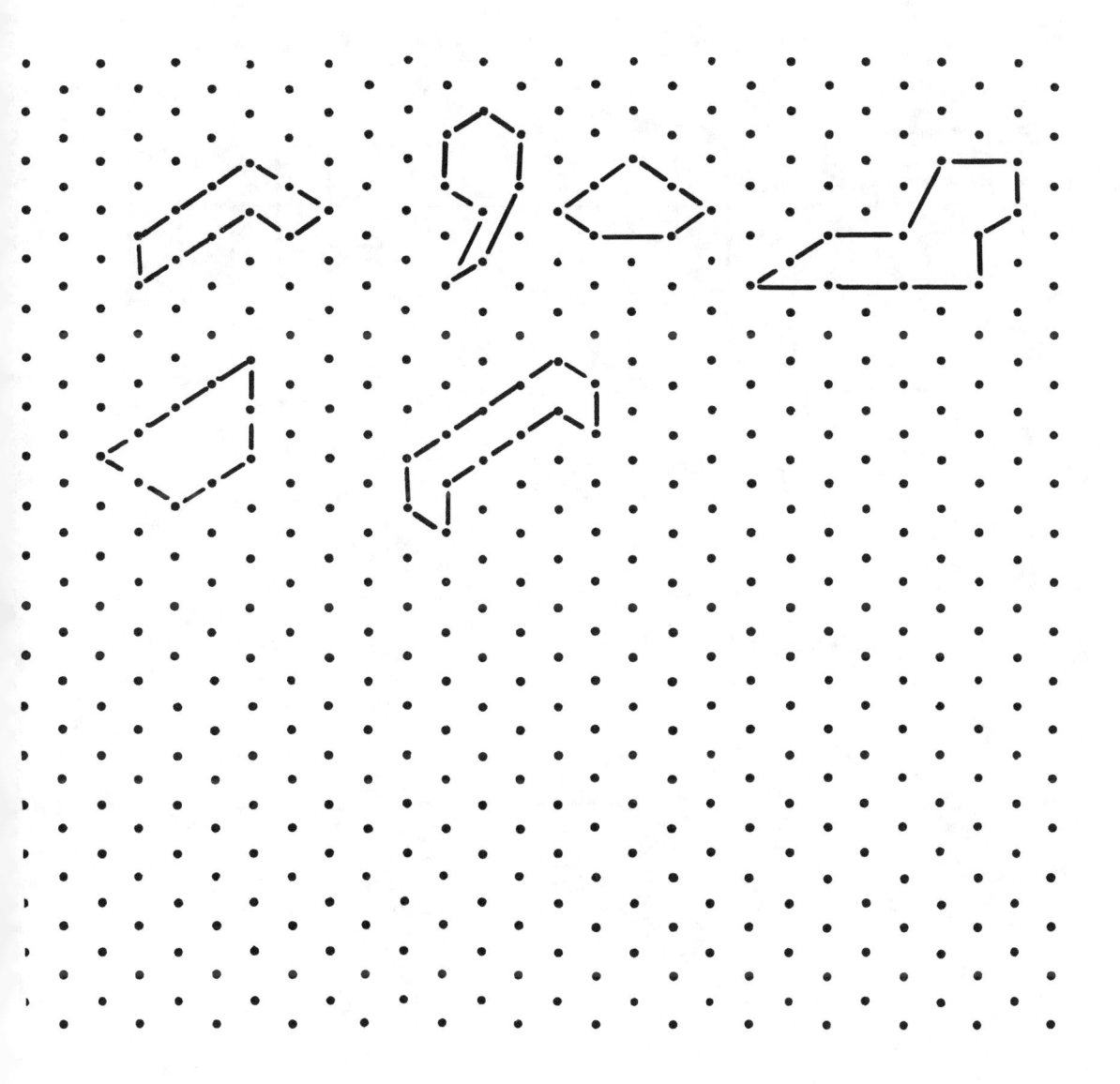

Horizontal and Vertical Plotting

Will any one piece nest to form a continuing pattern? Will one or more pieces nest to form a continuing pattern?

Shapes and Things

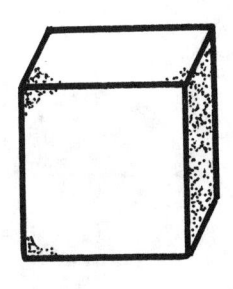

*3 faces - 8
2 faces - 12
1 face - 6
unpainted - 1

A wooden cube is painted black on all faces. It is then cut into 27 equal smaller cubes. How many of the smaller cubes are found to be painted on three faces, two faces, one face, and no face.

A logger cuts a pile of wood 8 feet long, 8 feet wide, and 8 feet high in exactly 8 days. How long would it take him to cut a pile of wood 4 feet long, 4 feet wide, and 4 feet high?

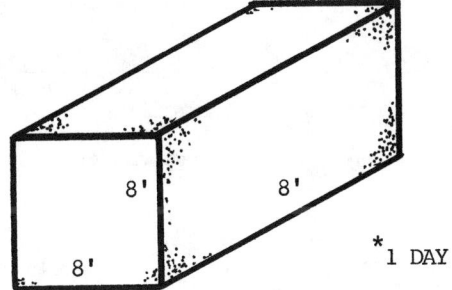

*1 DAY

Arrange these coins so there are two rows of four at right angles to each other.

*TWO STACKED COINS

Arrange ten houses in five rows of four houses.

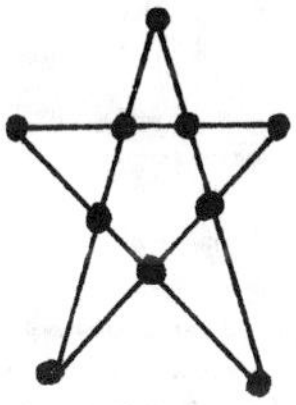

By using three cuts, create eight equal pieces of birthday cake.

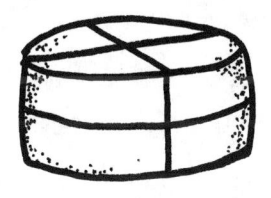

Can you divide this circle into eight sections by drawing just three lines? The lines do not have to be straight. The sections do not have to be of equal size.

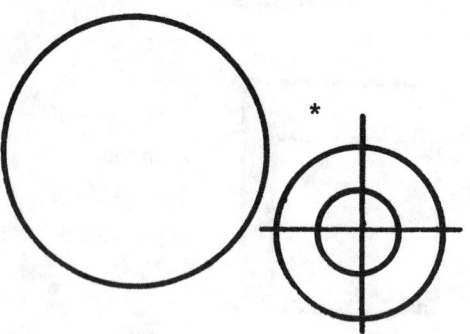

Three houses need the service of three utility companies - gas, electric, and water. These services must be delivered to each house but there are some restrictions on the method of delivery. No company's service maybe delivered if it goes over, under, or through another company's service lines or over, under, or through any other house other than the one being serviced.

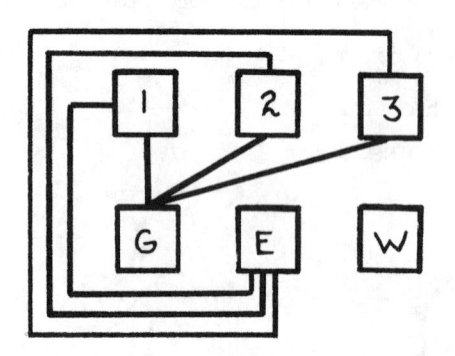

* Clues

Do not limit yourself to planar projections. Think three dimensionally. House or utility buildings may be shifted in any and all directions.

* The solution is said to lie in the area of rubber-sheet geometry or topology wherein the houses and the utilities buildings are positioned on a doughnut-like, inner-tube shape with the service being delivered in a spiral fashion about the exterior of the inner tube and spinning off to each house. This is supposedly accomplished well within the constraints of the problem.

* Can this be a solution?

    Use gumdrops and toothpicks.

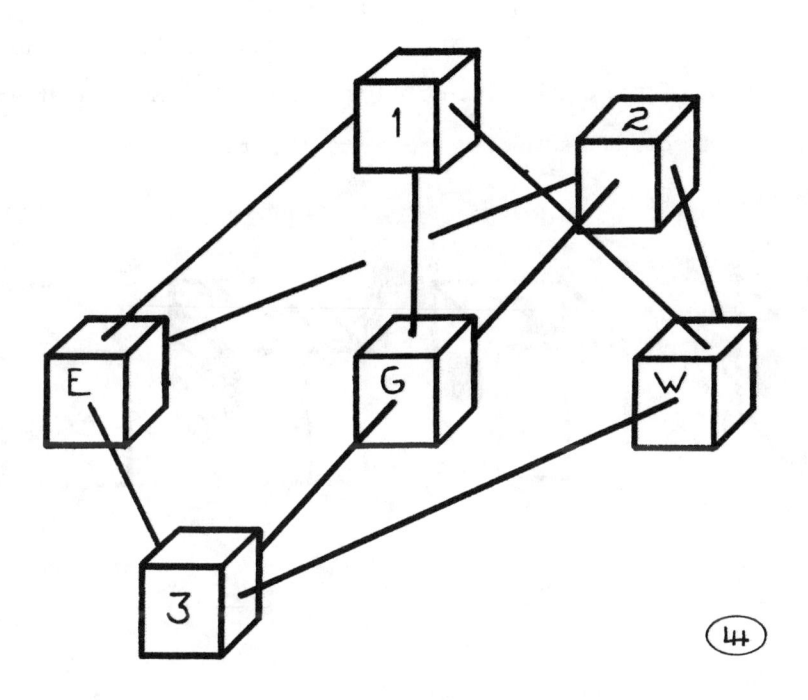

(44)

Is the piece of pie really missing?

* Look at it upside down.

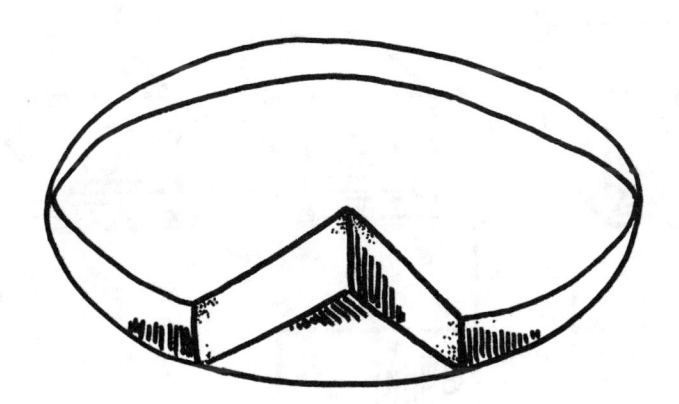

. Cubes with Six Different Face Patterns

Every cube is differently marked on each of its six faces (although all cubes have the same markings, but not necessarily in the same orientation, making one of the cubes a mis-match in the set). Imagine that you can pick up the cubes and turn them around to compare them. Can you find the odd cube?

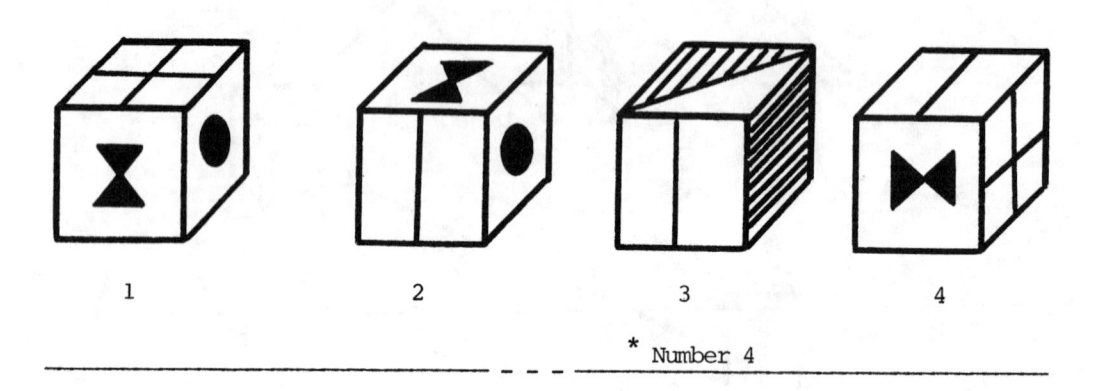

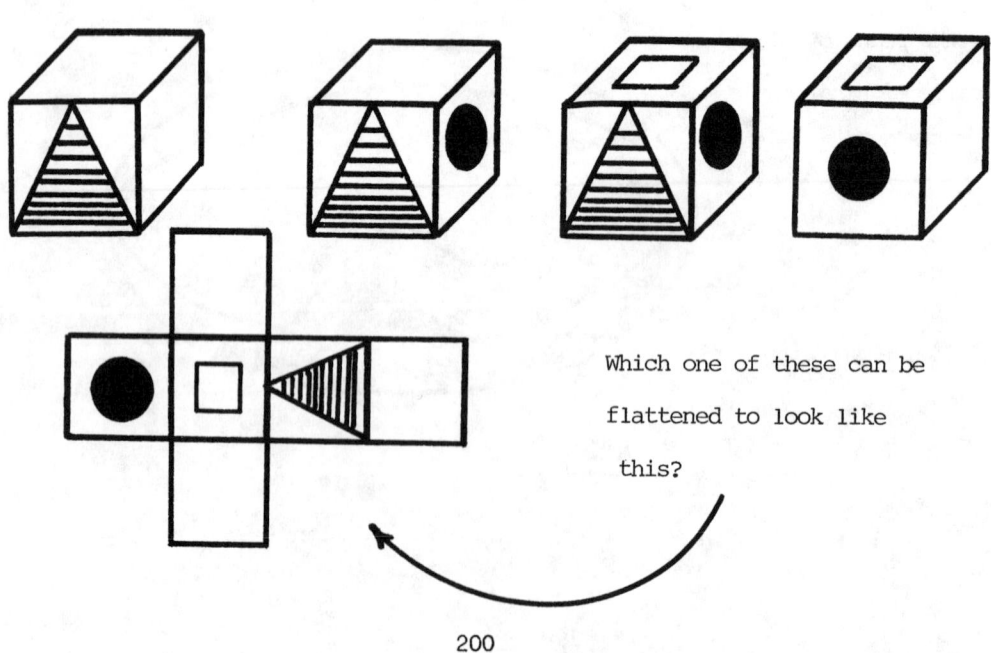

1   2   3   4

* Number 4

Which one of these can be flattened to look like this?

. Each three-dimensional figure corresponds to the two-dimensional layout accompanying it. The letters A and B are keyed to each other in both figures. Observe these figures. Match the numbers in one sketch with the letters for the corresponding edge in the other sketch (or vise versa). Use the letter "X" for those edges that do not show. If students experience some difficulty with the abstraction, have them construct and label the actual three-dimensional models to assist them in matching the numbers in the two-dimensional layouts.

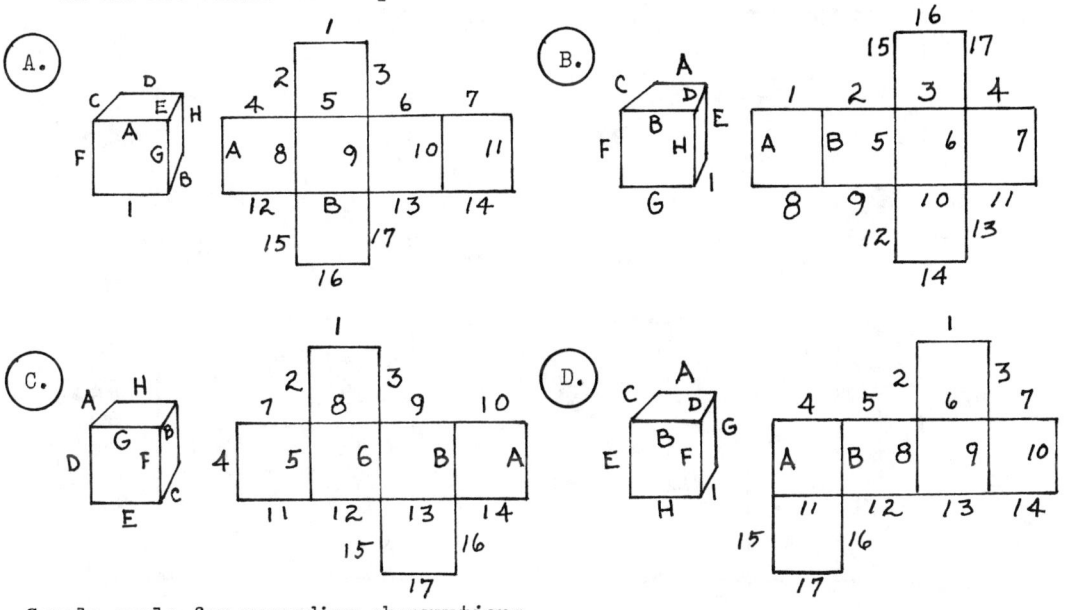

Sample scale for recording observations.

	1	2	3	4	5	6	7	8	9	10	11	12	13	14	15	16	17
A.	C	F	X	F	X	X	C	I	X	D	A	G	H	E	G	E	H
B.	C	F	X	X	G	X	A	D	H	I	E	H	E	D	F	C	X
C.	H	X	I	A	X	C	X	X	I	H	D	E	F	G	E	G	D
D.	C	E	X	C	E	X	X	H	X	A	D	F	I	G	G	F	I

201

. Observe each sequence both horizontally and vertically. Establish the pattern. Fill the vacant space in each sequence with the design you feel is appropriate.

A - F...easy to difficult

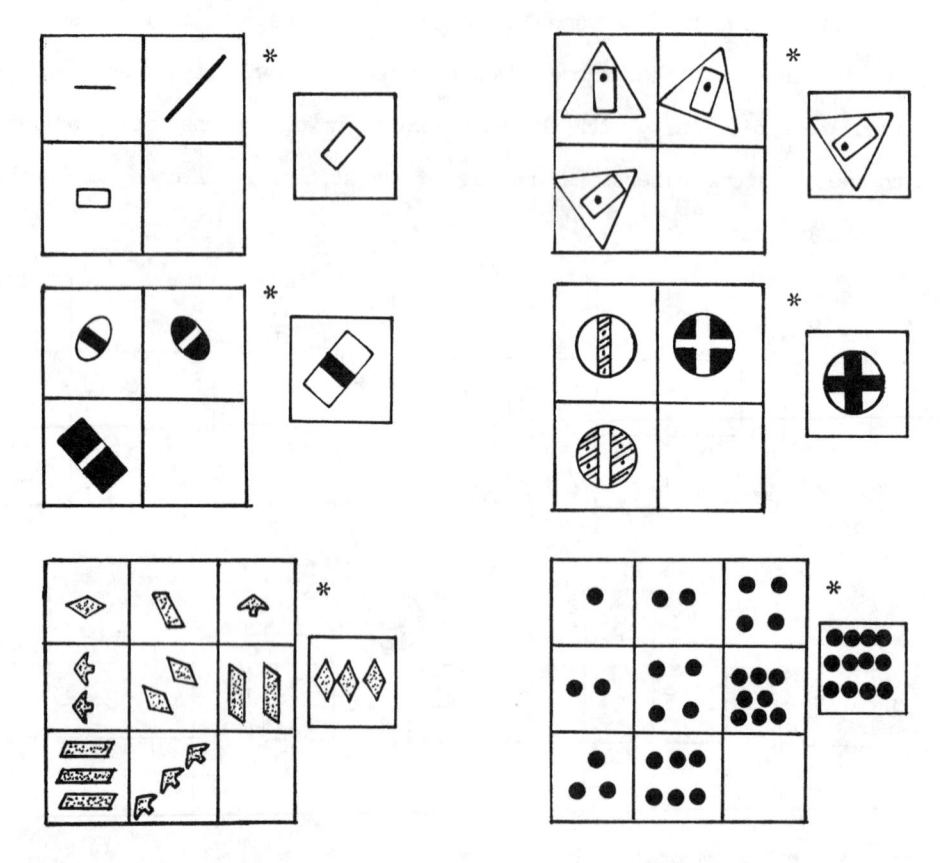

. How many squares do you see?

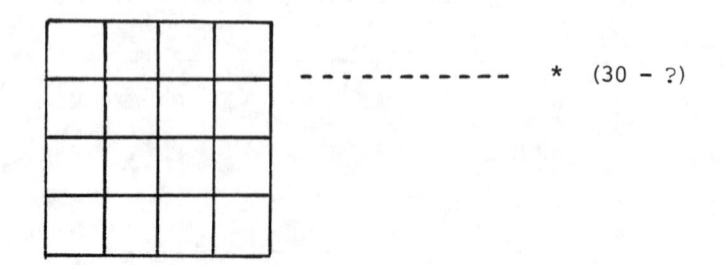

- - - - - - - - - -   * (30 - ?)

ORAGAMI - The extra special, spatial skill

Oragami is the Japanese art of folding paper. This art form serves many purposes. It has entertainment, therapeutic, and dexterity value. It is also extremely useful in demonstrating some principles of geometry. And, in the learning process, it is an aid to spatial development. Some samples are:

How to Construct a Drinking Cup (Easy).

<u>1</u> Use a square of paper. Fold it diagonally in half.

<u>2</u> Fold the right bottom edge to the left diagonal. Make a mark.

<u>3</u> Fold the bottom right corner up to the mark made previously.

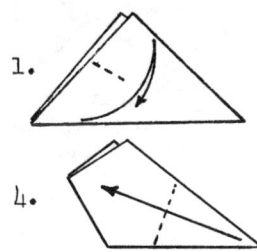

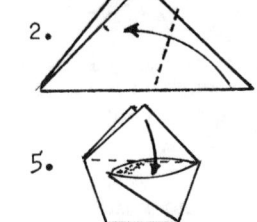

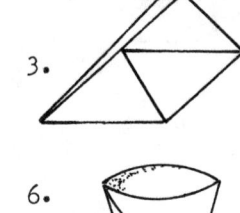

<u>4</u> Turn the paper over.

<u>5</u> Fold the bottom right corner over to the left.

<u>6</u> Tuck in the flaps to lock things in place.

How to Construct an Envelope (Easy). Use a rectangle of paper. Make a verticle center crease.

<u>1</u> Fold the top two corners to the center crease.

<u>2</u> Fold so that the edge meets the top point.

<u>3</u> Fold the left and right edges in to the center.

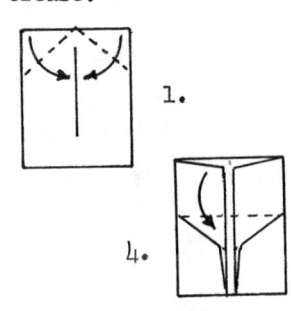

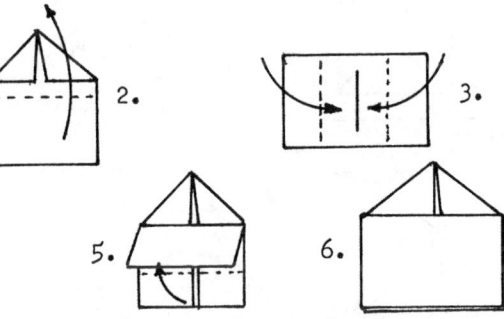

<u>4</u> Fold the top edge down as far as it will go.

<u>5</u> Fold the bottom edges up and tuck it into the pocket of the flap.

<u>6</u> Bring the top flap down.

How to Construct a Paper, fluttering butterfly (Easy).

Use a small square of paper.

1 Fold the top point to the bottom point.

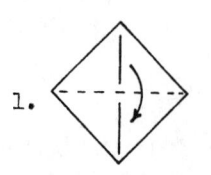

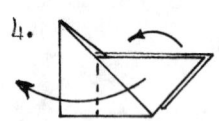

2 Fold the bottom point again so that it overlaps the top edge.

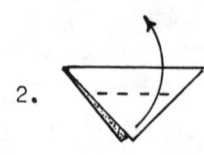

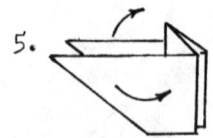

3 Fold the paper in half about the center crease. Left half behind the paper.

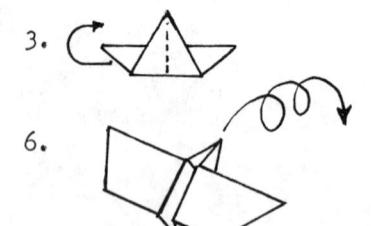

4 Fold the long flaps. The top one to the left and the lower one behind.

5 Raise the wings to a right-angle position to the body.

6 It is ready. Throw it, it should flutter to the ground.

How to Construct an Easy-to-open map fold (Medium).

Use a rectangle of paper. Mark the vertical center crease.

1 Fold the paper from the top to the bottom

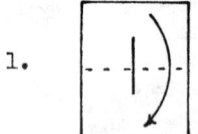

2 Fold the top corners down to the center crease and return.

3 Fold the top left corner to the front and open it out.

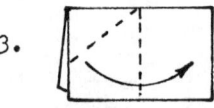

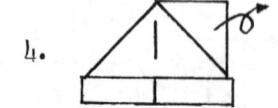

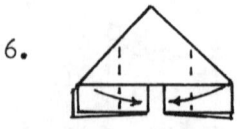

4 Turn the paper over.

5 Fold the left corner again and open it out. out.

6 Inside reverse fold the topmost left and right flaps so that the edges overlap at the center. Repeat this for the back side.

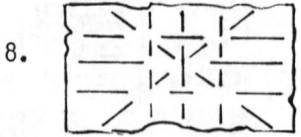

7 Completed map fold.

8 The opened map.

204

How to construct a dodecahedron using three, 8 inches x 10 inches sheets (Medium).

1. Cut the three sheets into 5" x 4" quarters.

2. Crease one rectangle vertically and horizontally.

3. Fold in two opposite corners so they meet in the middle.

4. Fold the other opposite corners to the center.

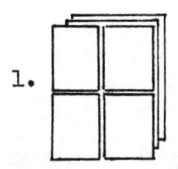

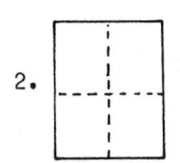

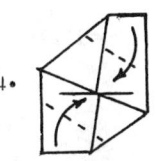

5. Fold the paper in half interlocking the two inside flaps.

6. Flatten

7. Fold point B in to the center line so that AB is parallel to CD.

8. Fold point E to the center line overlapping the left flap.

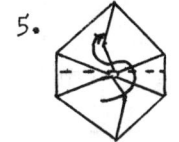

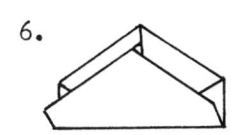

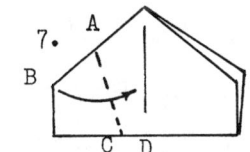

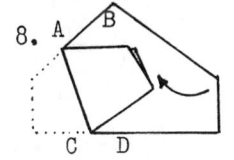

9. Then open two flaps.

10. This completes one of the units. Repeat steps 2 - 9 on the remaining rectangles.

11. Join three units by tucking the flap of one into the pocket of the other. Make four similar three-unit modules.

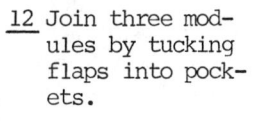

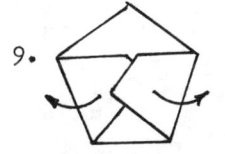

12. Join three modules by tucking flaps into pockets.

13. Add the last unit.

14. The completed dodecahedron.

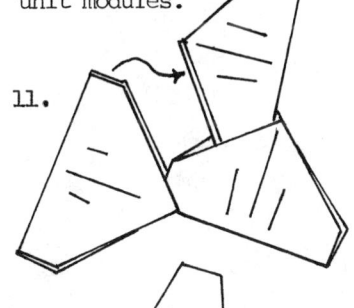

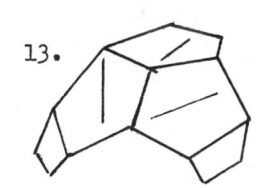

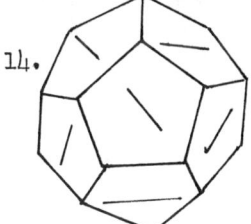

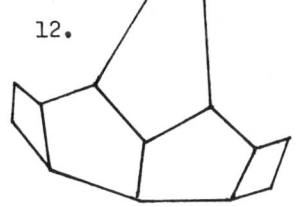

205

SPATIAL EXERCISE - NESTING STRUCTURES

. Construct a set of the following irregular shapes from 27 sugar, plastic, or small wooden cubes and white glue. Similar sets constructed by other participants are necessary to expand on the variety and complexities of additional geometric structures. These involvements integrate well with activities concerning surface area and volume.

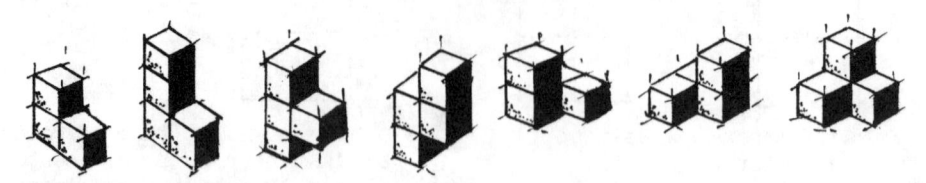

* The numerals associated with each construct indicate the number of pieces required for that structure. Sometimes the number of pieces must be drawn from one or more sets of the basic seven pieces. Two or more players can work together or they can compete with one another to see who arrives at the solution first. A good beginning point is to have the player attempt to arrange the seven pieces to form one large cube. This arrangement can take many forms providing the same end result...a cube!

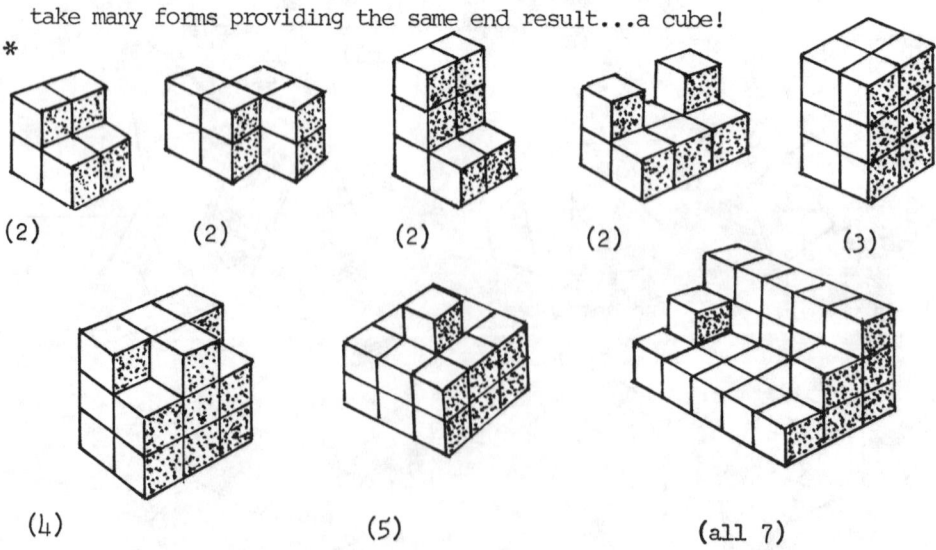

(2)   (2)   (2)   (2)   (3)

(4)   (5)   (all 7)

206

Spatial Blocks...calculating volume even when you cannot see.

The blocks are arranged in sets of nine. Progressing alphabetically within sets, the nine structures become increasingly complex. Examine each block structure. If there is a block or cube missing from the outside of the structure and you have difficulty seeing what is behind or below it, assume that all the blocks in that row or column are missing.

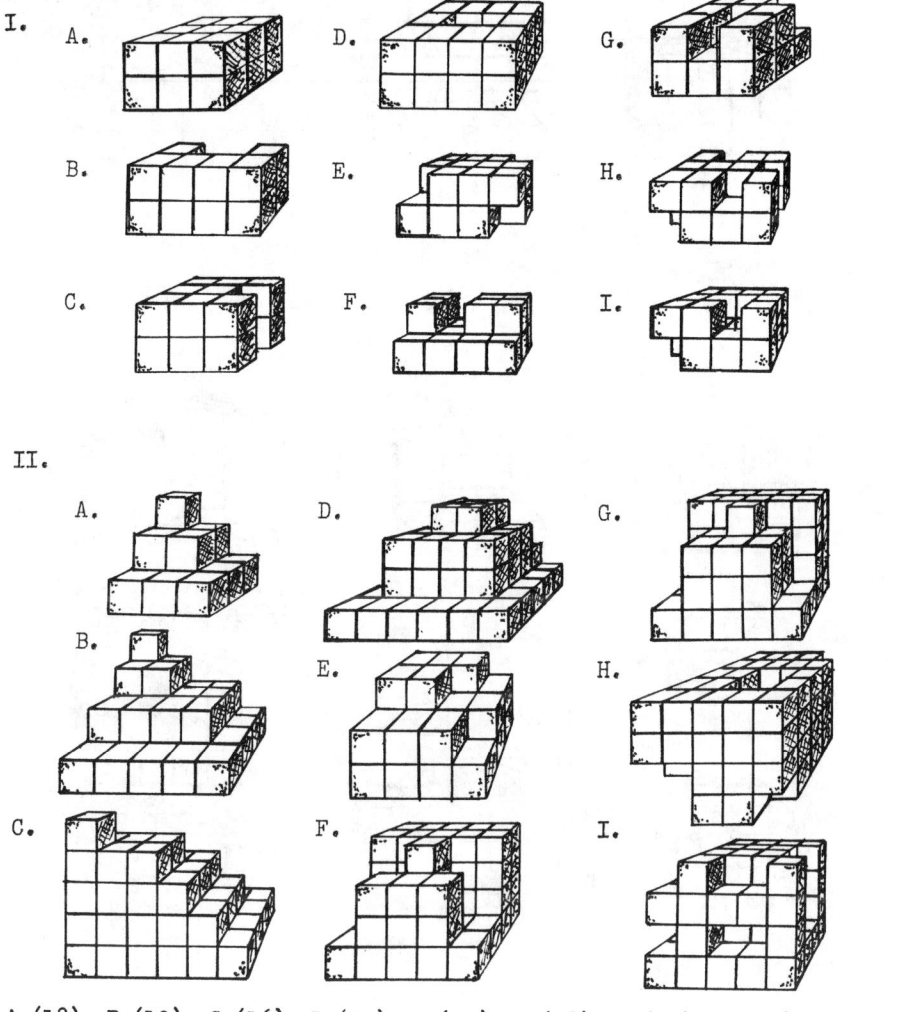

I.

A.   D.   G.

B.   E.   H.

C.   F.   I.

II.

A.   D.   G.

B.   E.   H.

C.   F.   I.

I.   A (18), B (12), C (16), D (20), E (20), F (18), G (15), H (18), I (18).
II. A (14), B (37), C (58), D (62), E (28), F (59), G (63), H (71), I (62).

III.

A.   D.   G.

B.   E.   H.

C.   F.   I.

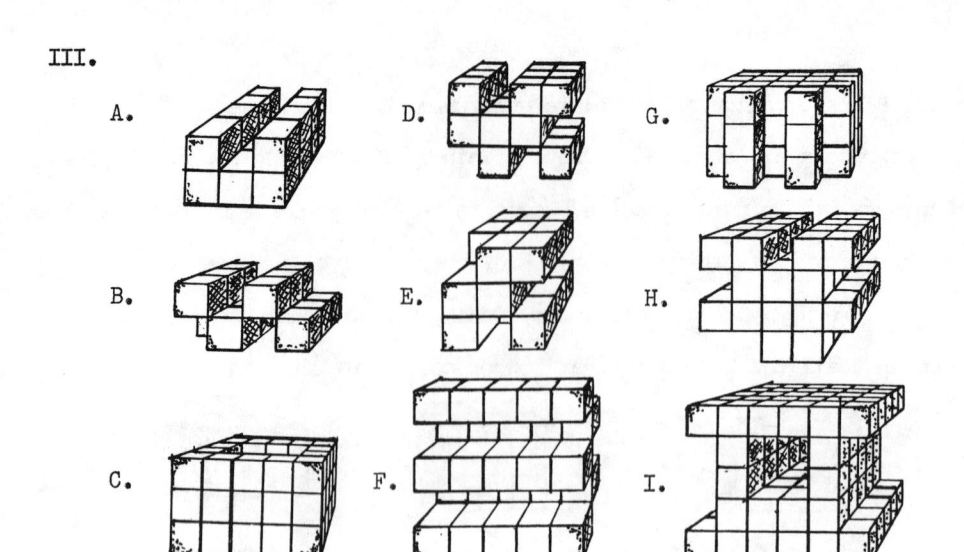

III.

A (20), B (14), C (42), D (25), E (22), F (25), G (48), H (56), and I (100).

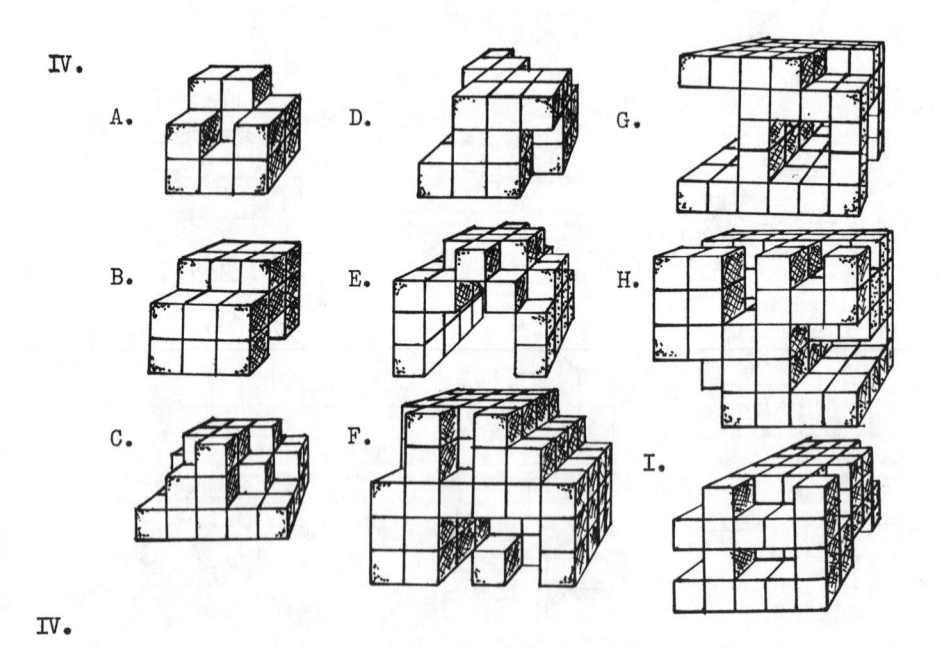

IV.

A.   D.   G.

B.   E.   H.

C.   F.   I.

IV.

A (13), B (21), C (41), D (31), E (38), F (93), G (70), H (93), and I (74).

Three Dimensional Plotting

Elementary school children are introduced to the number line concept in the early grades. This is usually presented as a horizontal line. By the intermediate school level, children are knowledgeable about planar grid systems and can do graphing and plotting using a two coordinate system. Few however, are familiar with polar coordinate systems. Many know what is meant by the expression "Where are you coming from?" but few know the meaning of a reference line or reference points. Expanding on the number line can lead to a fuller understanding of dimensionality be it linear, planar, or volumetric. Diagrammatically, this may be shown:

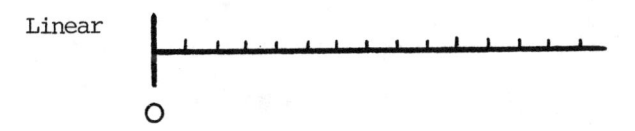

Linear

O

Swing the number line ninety degrees in a counterclockwise motion.

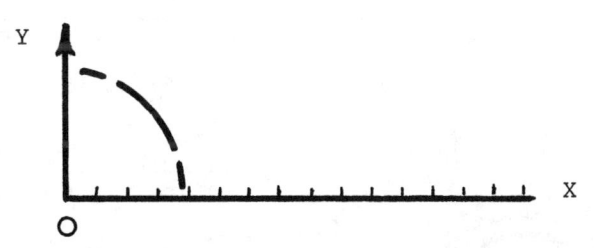

Y

X

O

This establishes a horizontal axis (X) and a vertical axis (Y). Their point of intersection is the origin (O). Uniform, number-line scales emanate in four directions in the same plane from this point of origin and describe four quadrants. Horizontal and vertical projections extended from appropriate, number-line scales provides one with a uniform grid.

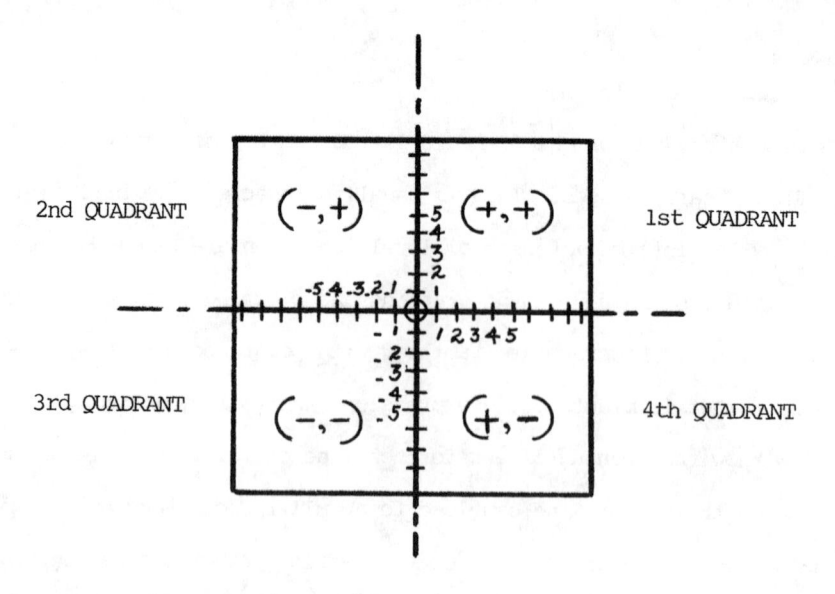

Using the standard technique for plotting points in a two coordinate system we can locate a position within the grid.

Polar coordinates are useful to describe a location within a circle as opposed to the X, Y coordinate, rectangular grid system. Again, the number line becomes essential.

Swing a complete arc about a fixed point, in this instance the "0" point. This generation will trace a circle. If an ink marker could be positioned at each graduation on your number-line scale so that when the number line was swung to form a full circle, you would trace a series of uniformily separated, concentric circles about the point of origin.

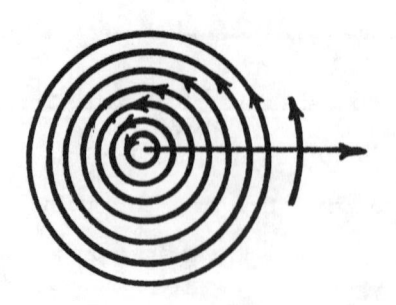

Combining this series of concentric circles with compass direction notations, plus their corresponding degree readings, one now could, by using the number line concept as measured from the center of origin and the degree reading, determine a location within a circle.

"There is a fly on the pie."

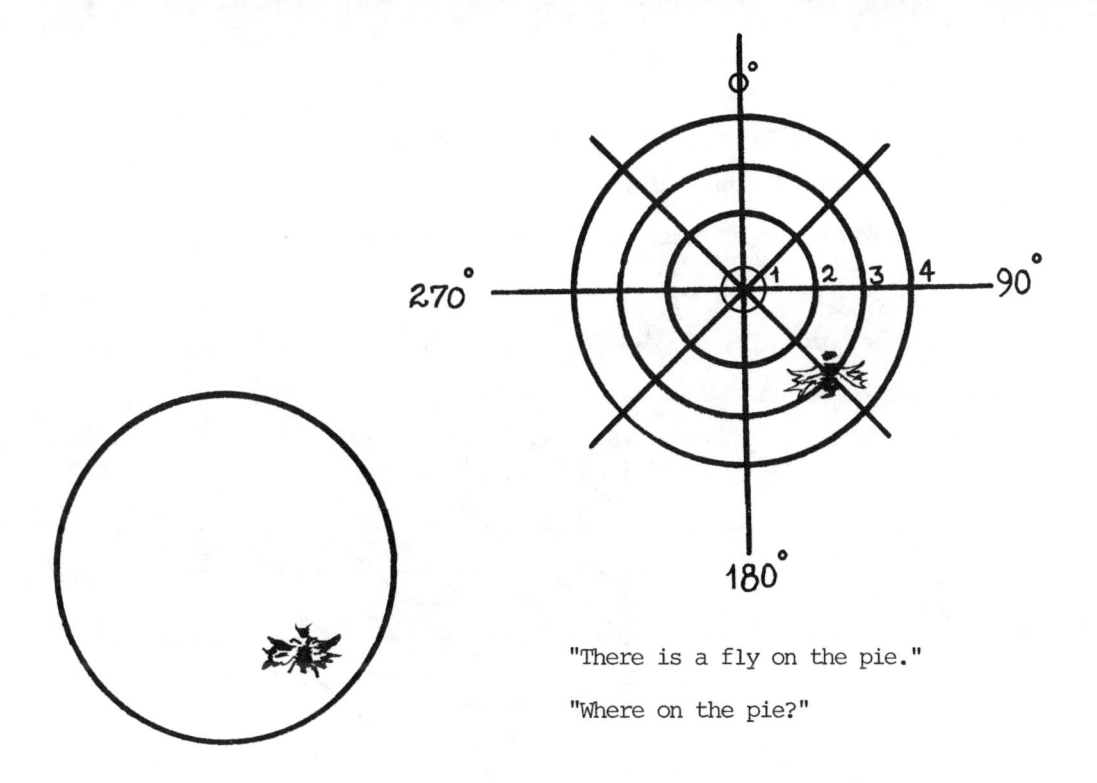

"Where on the pie?"

Again, the number line is needed for the generation of a three-dimensional, coordinate system.

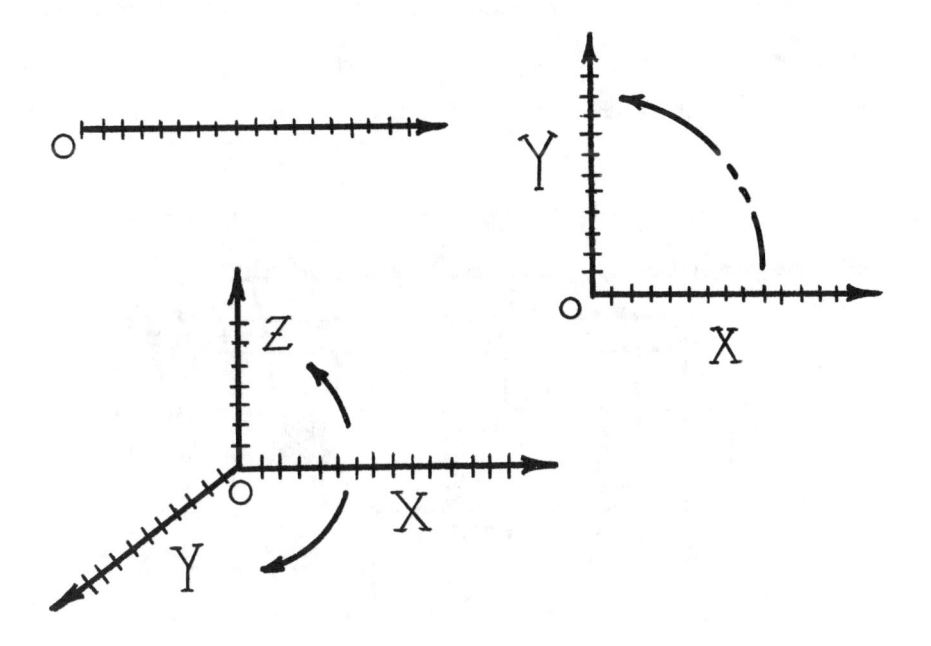

211

Swing the number line ninety degrees in a vertical plane to the original number line position. Then at right angles (ninety degrees) to this plane generate another number line. You now have three number lines existing at right angles to each other. These lines could be extended to include positive and negative, number-line scales. Establish which number lines will serve as which reference line for the X, Y, or Z axes.

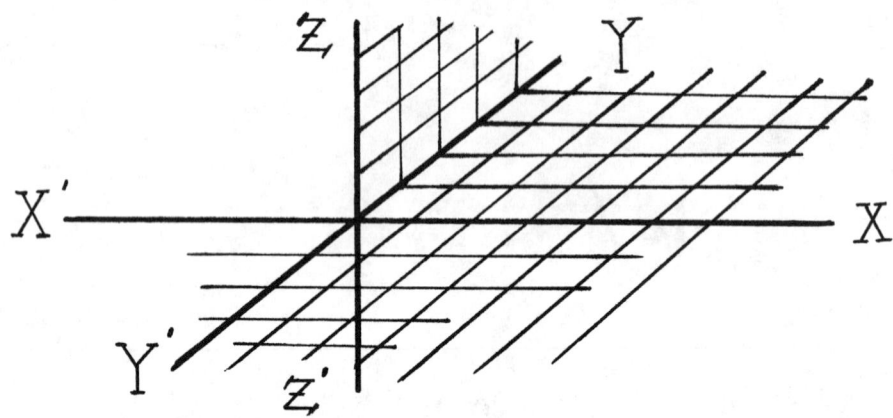

Small, plastic, grid-like containers are obtainable at local grocery stores. They usually are purchased containing berries, cherry tomatoes, etc. These make excellent devices to provide spatial experiences for children. Soda straws cut at unequal lengths, set in clay to hold them erect, are positioned within the containers. With reference lines established and agreed upon, children locate the positions of the tip of each straw in a three-dimensional space.

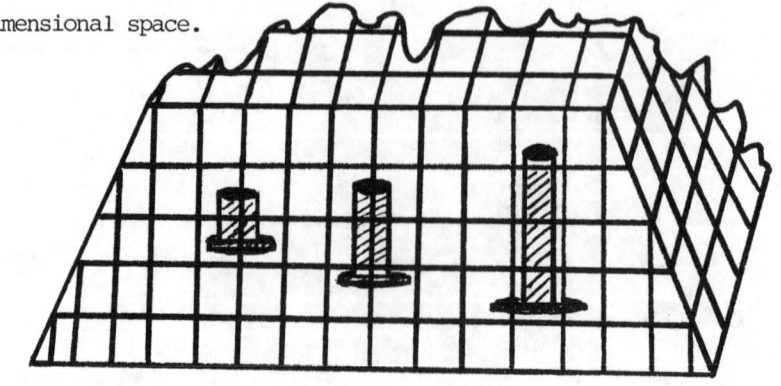

. Three Dimensional Projection

A cow is viewed differently when viewed from opposite ends. And, so it is with three dimensional projections. One view does not always do it. A cube is a cube, is a cube viewed from the top, the bottom, or any side view. This is also true for a sphere. But the same is not true for all geometric shapes or groups of geometric shapes.

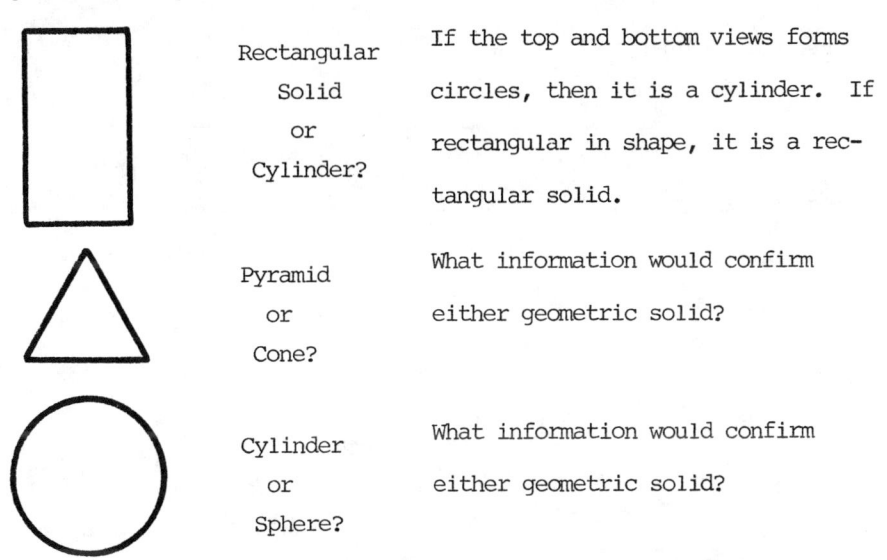

Rectangular Solid or Cylinder? If the top and bottom views forms circles, then it is a cylinder. If rectangular in shape, it is a rectangular solid.

Pyramid or Cone? What information would confirm either geometric solid?

Cylinder or Sphere? What information would confirm either geometric solid?

From observations of single geometric solids, move to observations of geometric solids in groups. Select two geometric solids. Arrange these as one.

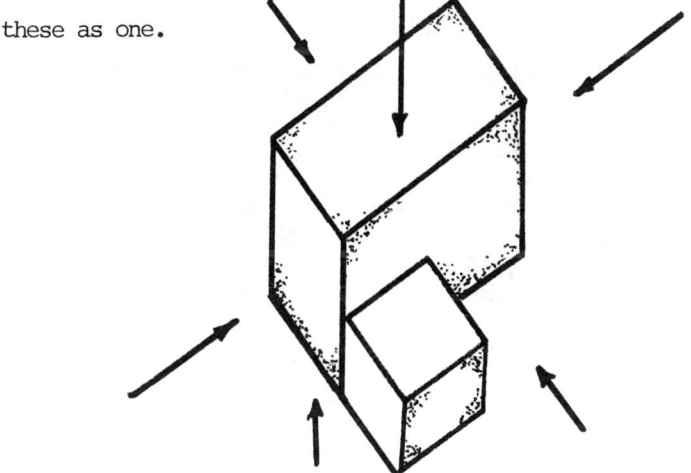

213

Construct from a Manila folder a cube (see six crystal systems for cubic cutout) whose dimensions are larger than the overall dimensions of the two pieces you have assembled. Cut six squares of paper with the same dimensions as any one side of the cube. Draw each of the six views, one on each of the squares. Glue the squares to the cube in their correct orientations. When dry, the cube can be opened and flattened to show a three-dimensional representation in a planar view.

Repeat this procedure utilizing three geometric solids; two of which are so arranged so as not to be seen in one view (A) and one piece which cannot be seen in alternate view (B). These disparities are to reveal to the viewer the necessity for "getting all the views" in order to see the total picture.

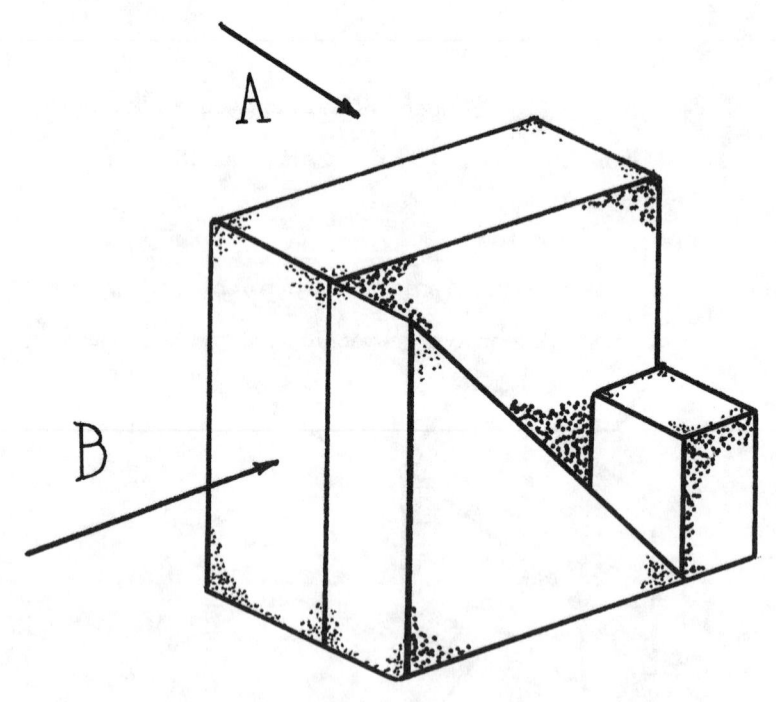

Six Crystal Systems

All crystals are readily classified into six groups called crystal systems. This classification can also be established by considering the lengths and angular relationships of imaginary lines, passed through the center of a crystal called a crystal axes.

- Cubic System - Crystals in this system have three equal and perpendicular axes.

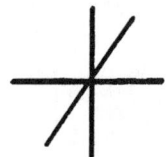

- Hexagonal System - Hexagonal crystals have four axes. Three are equal, horizontal in the same plane and intersect at 60 degrees. The fourth axis is perpendicular (vertical) to these. It can be longer or shorter than the horizontal axes.

- Tetragonal System - Crystals of this system have axes which intersect at right angles. The vertical axis is longer or shorter than the two equal horizontal axes.

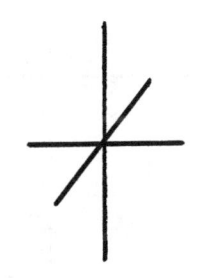

- Orthorhombic System - This system is recognized by three perpendicular and unequal axes.

- Monoclinic System - The three axes of this system are unequal. Two intersect at an oblique angle, and the third is perpendicular to them.

- Triclinic System - Triclinic crystals have three axes, all unequal, all inclined to each other.

CUBIC

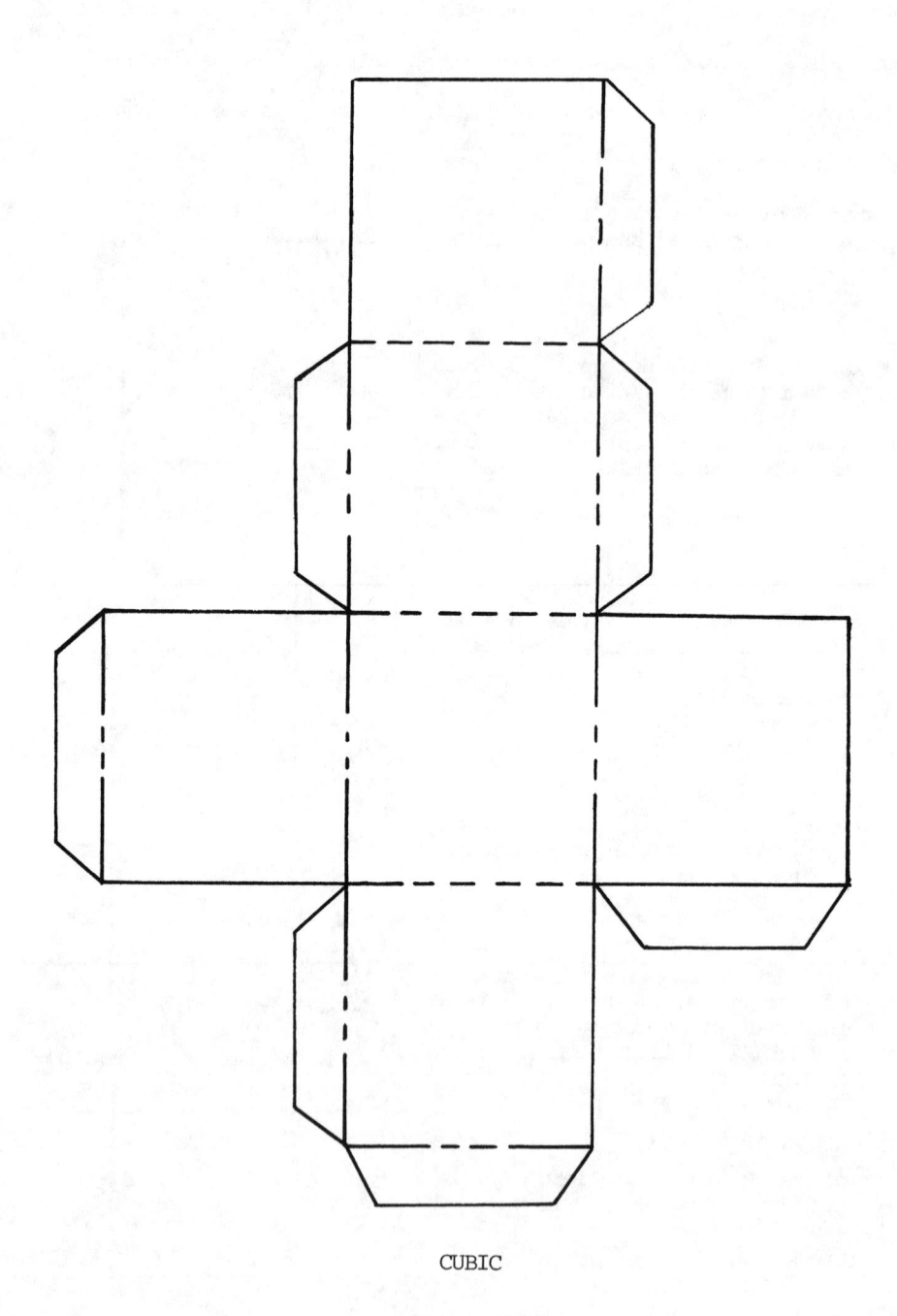

216

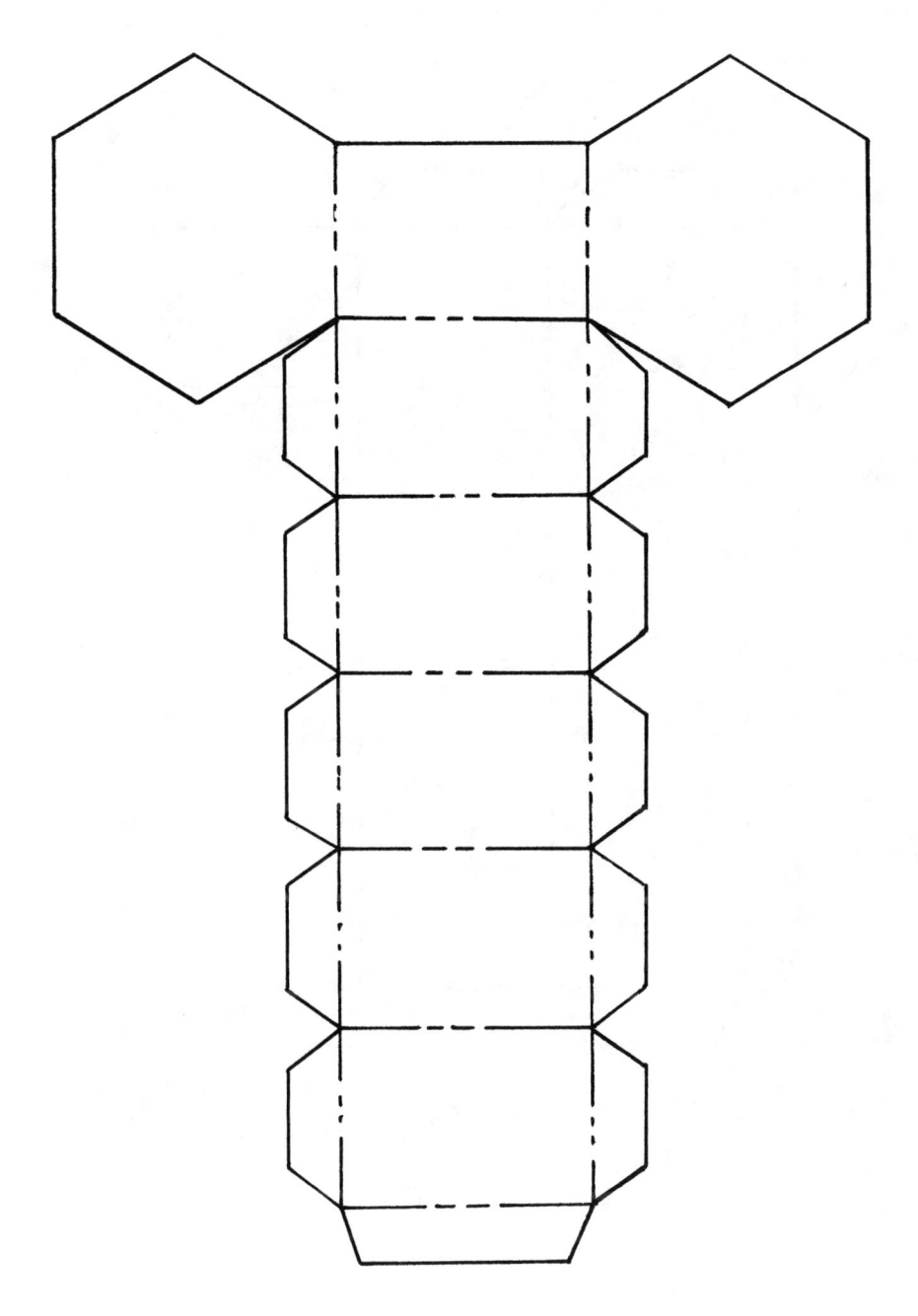

HEXAGONAL

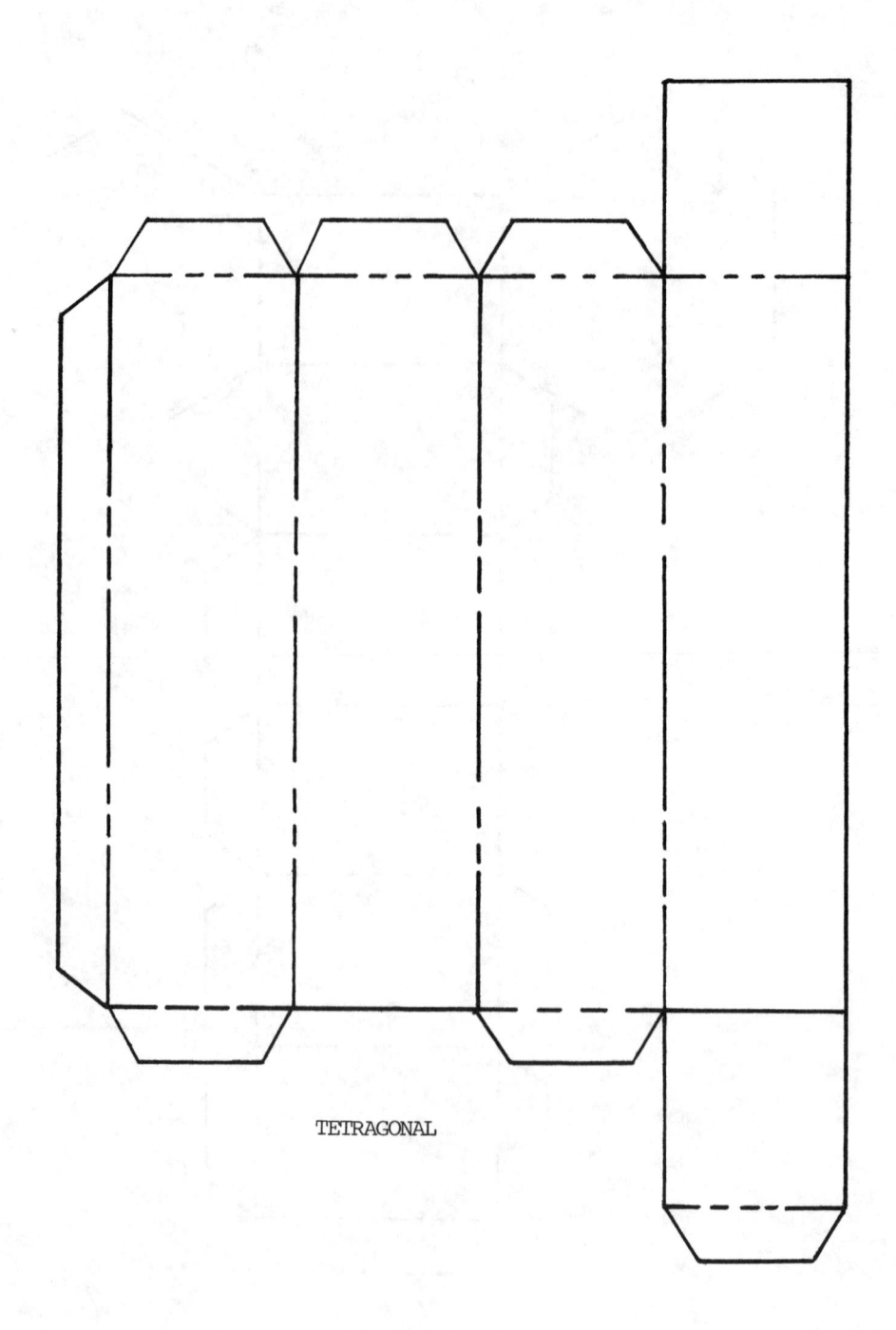

TETRAGONAL

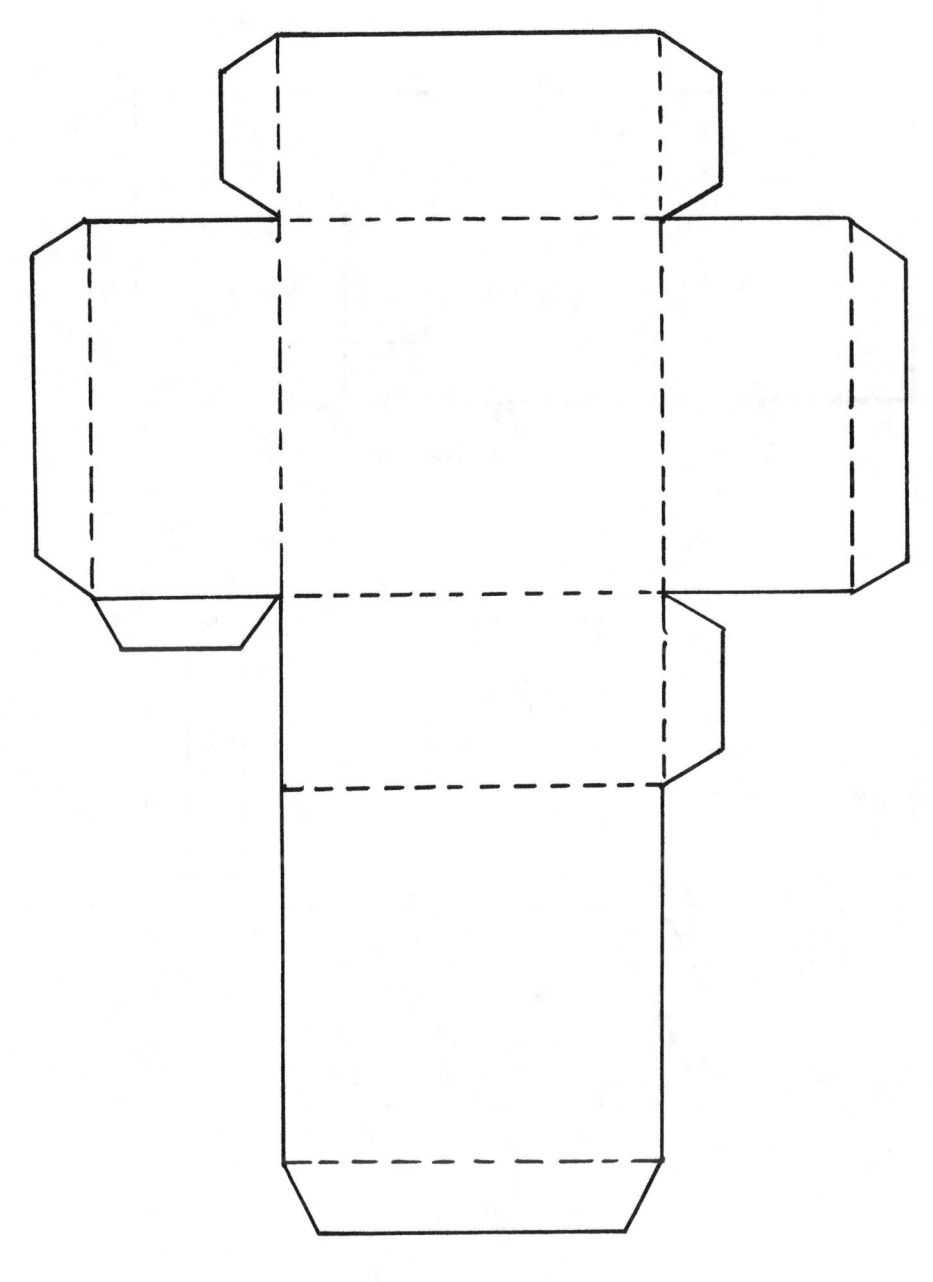

ORTHORHOMBIC

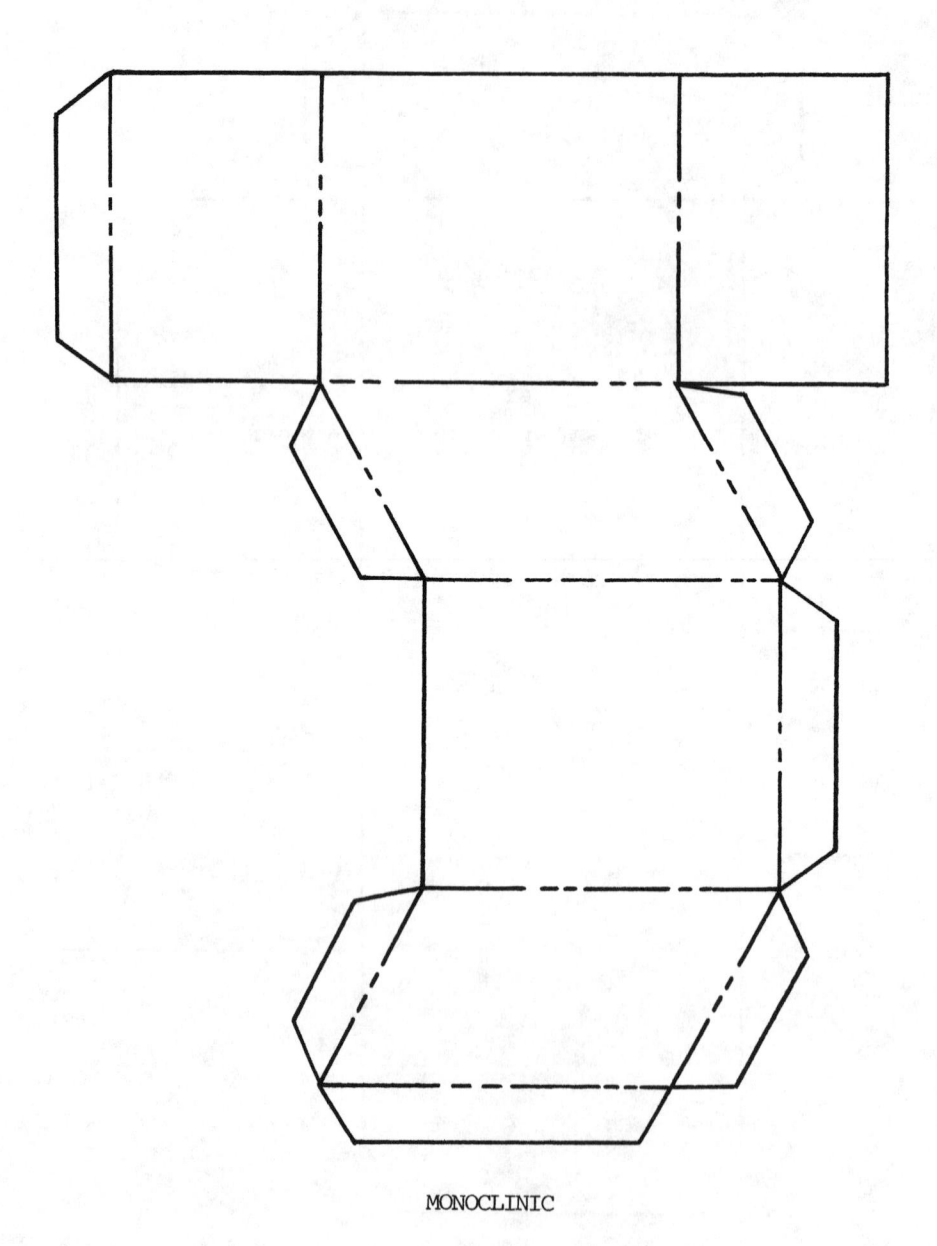

MONOCLINIC

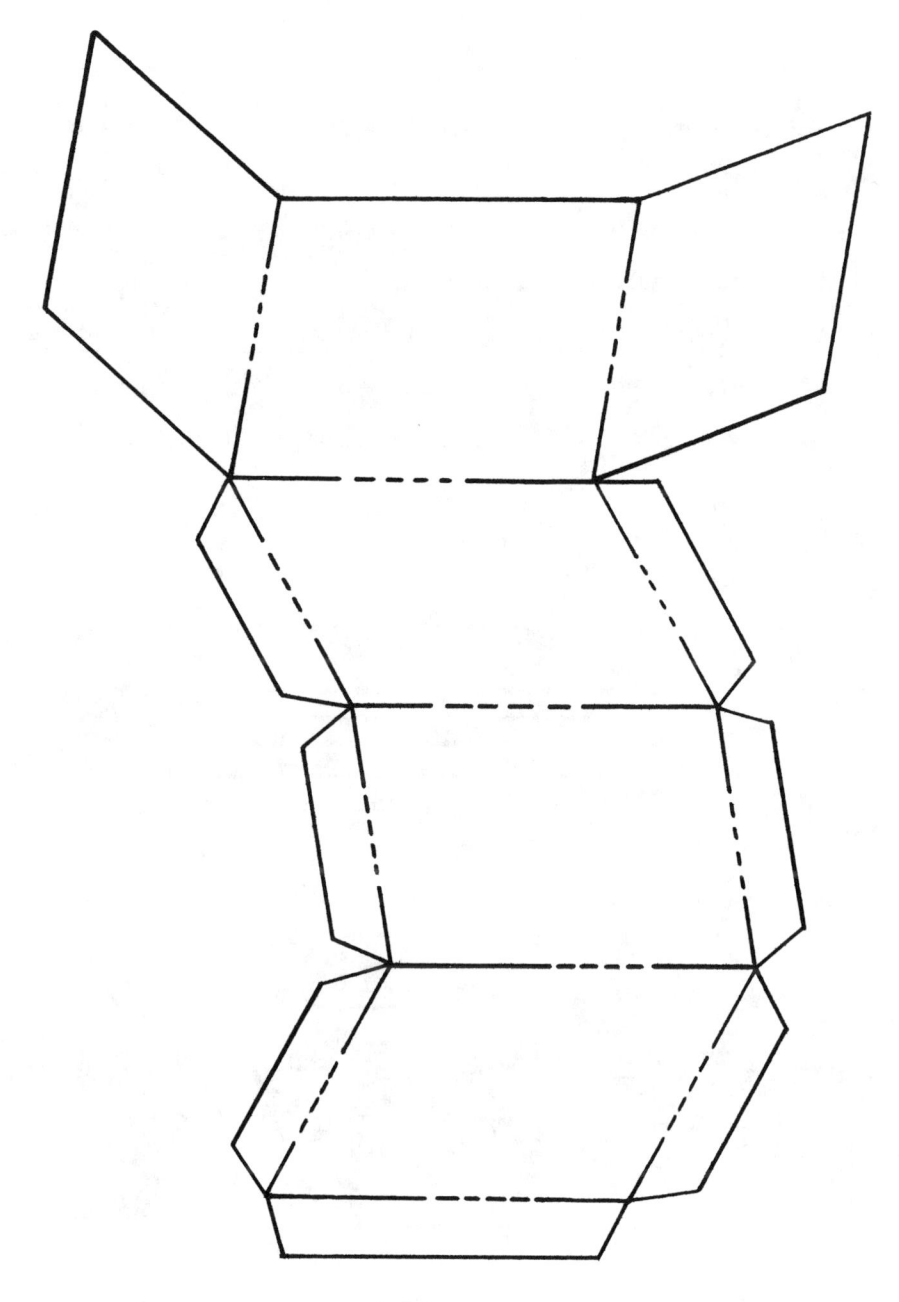

TRICLINIC

. Three Dimensional Cross-Sections

The area of geologic sciences abounds with situations which necessitates proficiency with spatial problems. A few are geologic faulting, folding of geologic structures, and subsequent erosion accompanying these phenomena.

A three-dimensional model can be easily constructed to depict normal faults, reverse faults, transverse faults, or any combination of these. These pieces should be selected for differences in color or end-of-piece grain patterns. These differences can be described in the model as different stratum, each with unique characteristics. Using wood glue, join these three pieces together. When the glue has thoroughly dried, sand the block and cut the block diagonally.

The fault model, by the appropriate gyrations, can be used to connote a variety of faults.

NORMAL FAULT   REVERSE FAULT   TRANSVERSE FAULT

Depending on the end-of-piece grain view selected, any one of the three boards used, could connote folding, synclinal or anticlinal folding, or simple layering.

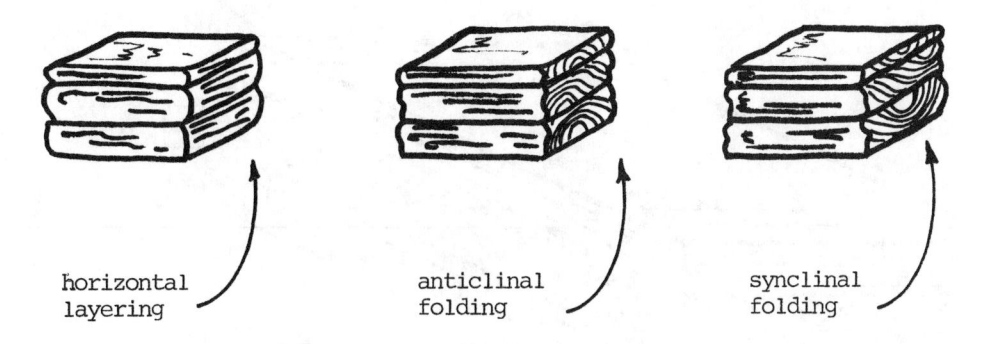

horizontal layering

anticlinal folding

synclinal folding

Demarcations between glued pieces of wood can be viewed as hiatuses wherein inferences can be made as to erosional occurences that might have occurred in the interim depicted by difference in geologic structures comparing one board segment to another. Depending on the orientation of the components of the constructed fault mode, various sequences of events can be postulated.

Three Dimensional Subsurface Geologic Models

Construct from a Manila folder a series of 3" x 5" x 2" rectangular solids (see six crystal systems for orthorhombic cutout). Graduate the classroom involvements from simple to complex analytical situations.

Which of the following events are visible in the geologic models and which, if any, occurred first, second, third, etc.

faulting

tilting of beds

beds deposited horizontally

an intrusion into pre-existing material

horizontal beds folded

* Clue: Sedimentary beds are formed in horizontal layers

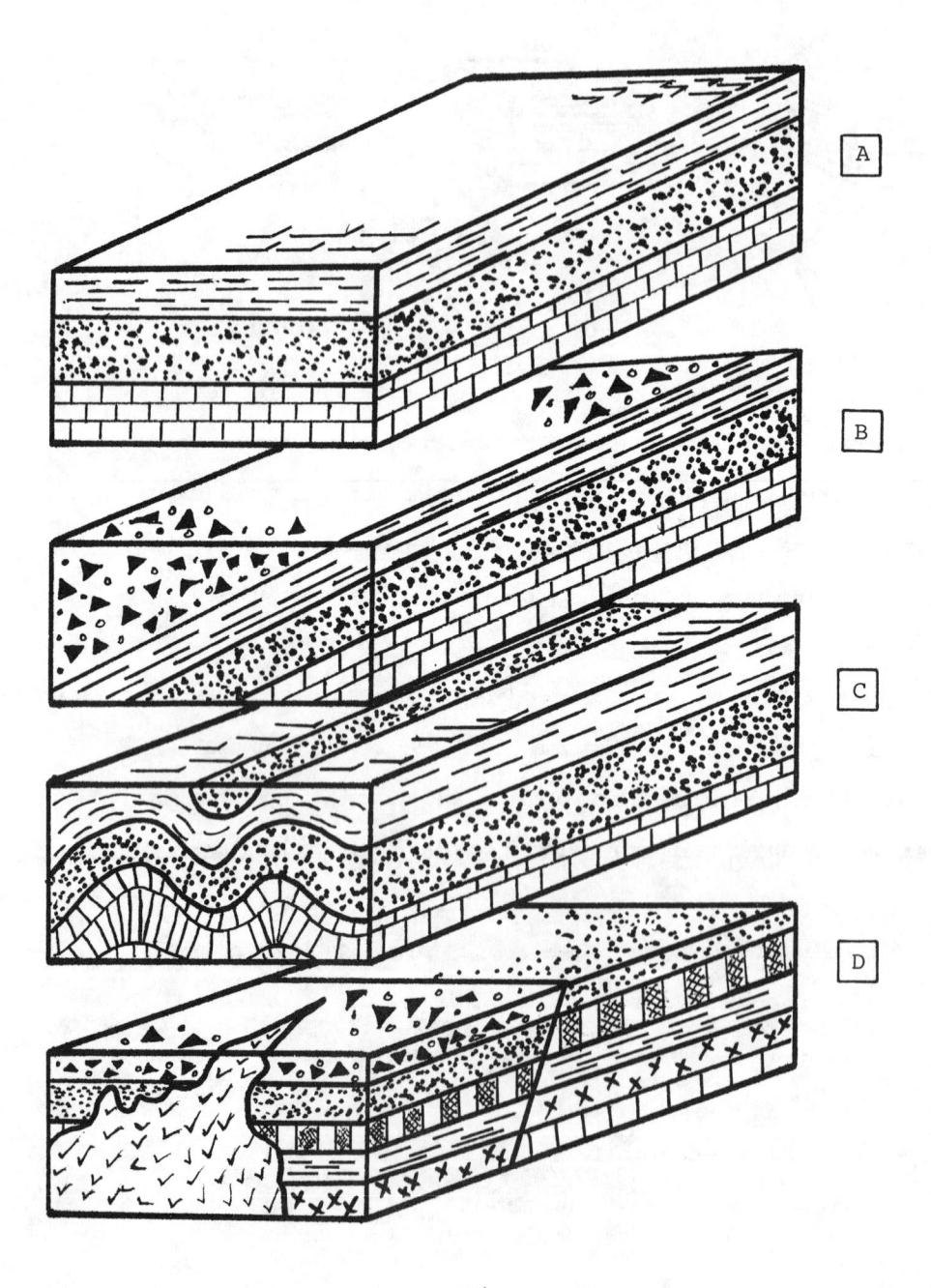

A

B

C

D

A simple but interesting problem which simulates subsurface geologic mapping of subsurface structures is:

There is evidence that an iron ore body lies below a portion of a broad flat valley. This area is 1/4 mile square. Map on the surface of the ground a grid (horizontal and vertical lines intersecting at right angles). At each point of intersection a drilling rig will be erected. Drilling will continue until the ore body (if it exist) is reached. A picture of the surface of the land and the grid looks like this:

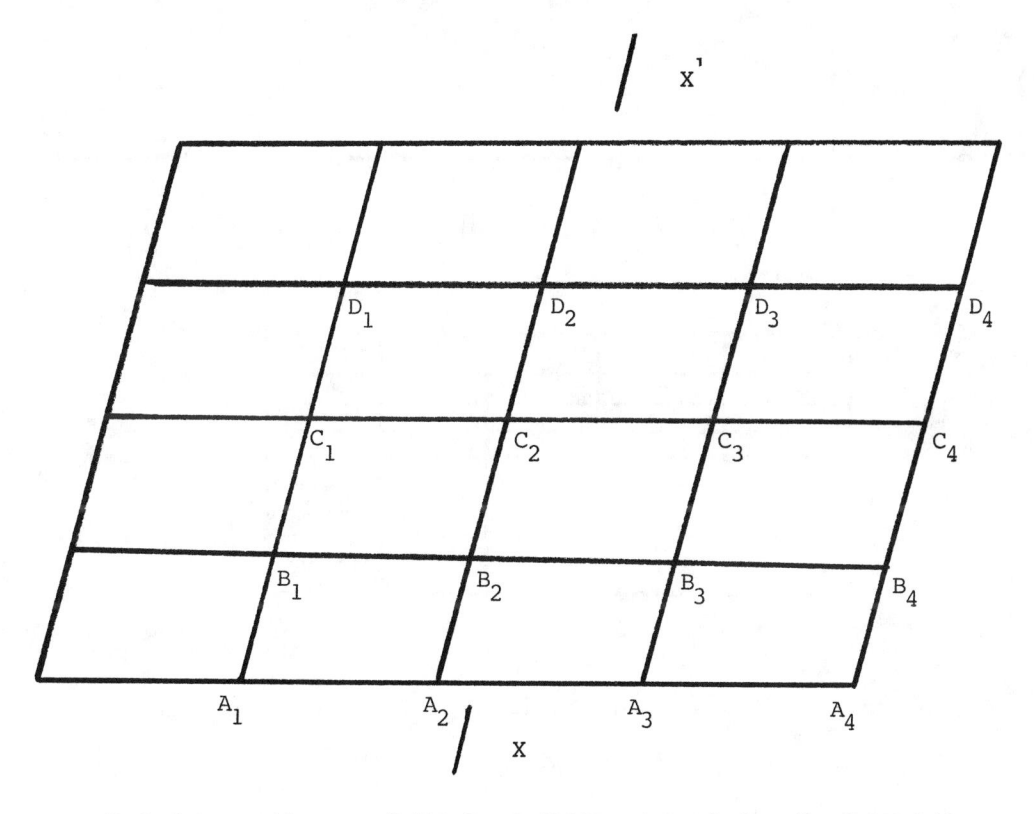

Each intersection was drilled. Drilling stopped when the drill bit revealed that the iron ore was reached. The data was recorded. The valley floor is at sea level. The accumulated data reads as follows:

Drill Site	Ore reached at a depth of	Drill Site	Ore reached at a depth of
A_1	50'	C_1	40'
A_2	60'	C_2	50'
A_3	70'	C_3	60'
A_4	80'	C_4	70'
B_1	40'	D_1	50'
B_2	50'	D_2	60'
B_3	60'	D_3	70'
B_4	70'	D_4	80'

A cross section at X -X' would give you a picture like this:

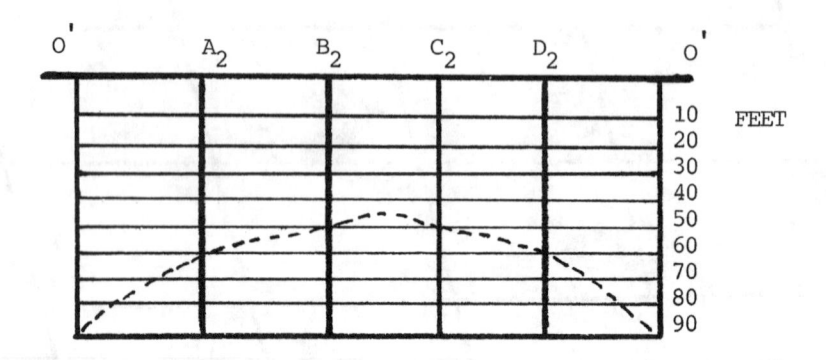

Draw or model in clay, using the data furnished from the 16 wells, a cross-section view. What shape will the subsurface ore body trace?

. Rubic Cube

The rubic cube is an excellent device to stimulate transverse rotations and to advance or retreat differently colored squares from one position to another.

ON TEACHING

POUR IN EQUAL PARTS OF ANTICIPATION AND PREPARATION;

ADD EIGHT HOURS SLEEP FOR THE TEACHER, BEFORE

TEACHING;

SEASON WITH A SENSE OF HUMOR;

AND USE SUFFICIENT RESOURCEFULNESS TO KEEP THE MIXTURE

FROM BECOMING LUMPY!

Section IV: Discrepant Events, Puzzlers and Problems, and Tenacious Think Abouts

DISCREPANT EVENTS AND SCIENCE INSTRUCTION

Discrepant events are events which when they occur the results are contrary to what an observer might have originally anticipated. Some educators call these events <u>mind captures</u>. Some called them <u>contretemps</u>. Others call them <u>unexpected happenings</u>. By whatever name they are called, discrepant events are excellent devices to stimulate interest in the learning of science concepts and principles.

Discrepant events usually are used to introduce a lesson. However, many can be expanded beyond just an introductory item by applying either the morphological, process, or ideation-generation approach to the presentation. Basic to usage of discrepant events is the presentation of the material in a somewhat theatrical manner either by the manipulations of the material or by the nature of the discrepant event itself. Excitement is generated by the presentation of a unique phenomenon and an unexpected contradictory result.

Not every lesson in science can be preceded by a stimulating, captivating, discrepant event. Some areas of science do not lend themselves to a sprite introduction through use of discrepant events. However, where applicable, they illuminate science instruction.

Expansion of a Solid and a Liquid

Fill a glass bottle to the brim with cold water. Fit the bottle with a one-hole stopper through which a length of glass tubing has been passed. Inserting the stopper into the glass bottle will result in some water being forced up into the tube where it remains. Place this entire apparatus in a can of hot (not boiling) water. As the water warms in the bottle it is expected that the water would be pushed up into the tube. At first, the water level in the tube drops. Then, it rises. How can this be explained?

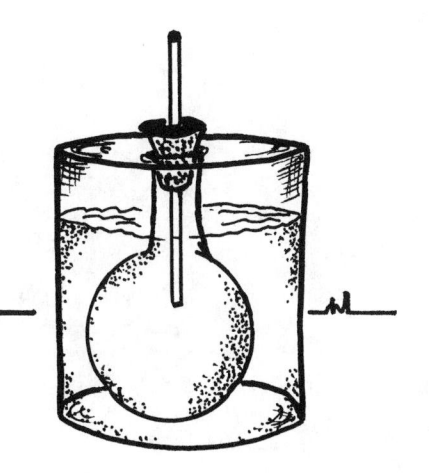

* When the glass bottle is placed in the hot water, the bottle becomes warm before its contents, so the glass begins to expand before the entrapped water. Thus, there is an increase in the bottle's capacity which permits some water to descend from the tube and cause the water level to fall. Later, when the cold water became warm and expanded, it pushed the water level up inside the glass tubing.

Energy

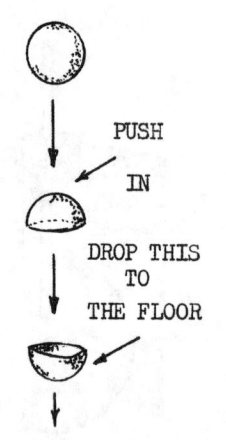

Using an old racketball, carefully and smoothly cut off about the top third of the racketball. This can be a bit fussy and if too much is either retained or removed, it does not provide satisfactory results. Do not discourage, collect a number of used racketballs and keep trying until desirable results are achieved. Be careful when cutting into the racketball.

Using the top third, press in on the top portion so that the inside of the ball becomes the outside of the ball in an inverted cup-like shape. Hold this object at about shoulder height and let it fall to the ground. This object will rebound in a surprising manner.

   * When the one third of the racketball is inverted, it is placed in a tension situation. When the object is dropped and strikes the floor, the object reverts to its original shape, releasing the stored tension, and rebounding vigorously.

Energy

Drop a ping-pong ball. Measure and record the height of its rebound. Drop a golf ball in the same manner. Measure and record the height of its rebound. Which object rebounded to the greatest height? Place the ping-ping ball vertically on top of the golf ball. Drop the two objects simultaneously. What do you observe?

   * The ping-pong ball will rebound much higher than it previously did. The energy of the rebounding golf ball is transferred to the ping-pong ball and it ricochets in an impressive manner.

Molecular Structure

Fill a pint jar about one-half full of corn starch. Add just enough water to make a thick paste. Ask children to slowly push their fingers directly into the paste. This will present no problem as their fingers will go in easily and fully. Follow this by inviting the children to rapidly punch their fingers into the beaker. They will be surprised that this time the material offers a high resistance and their fingers will not readily penetrate. Request explanations for their observations.

* The molecular structure of the starch-water material maintains its structural integrity on impact, but under slow but steady pressure it will give easily.

Temperature/Air Circulation

Place a jar of frozen fruit juice (or water) on a counter top to thaw. Later the jar was observed to have appeared to have thawed from the top down. How can this be explained?

* The frozen liquid container as it cools chills the surrounding air. This air being more dense than the room temperature air, sinks to the lower portions of the container. The warm air that replaces the sinking cooler air thaws the upper portion of the jar first.

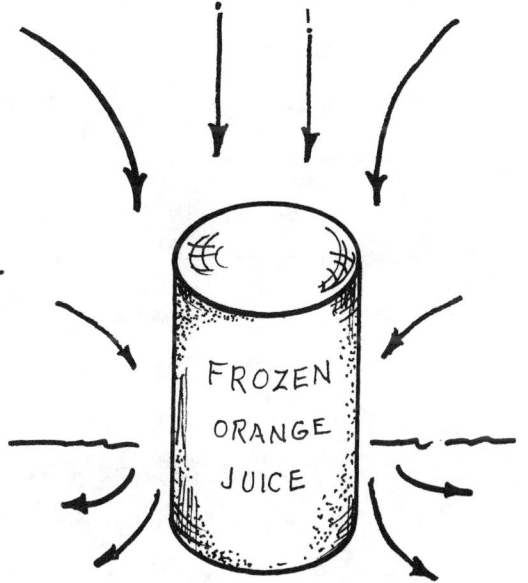

Air Pressure

Fill a glass tumbler or jar with as much water as it will hold. Place a piece of stiff cardboard (a 3 x 5 card will do) on top of the tumbler. Lay the palm of your hand across the card. Position this over a sink. Now invert the glass of water holding the card in place using your palm and closed fingers. Do not let go of the card until you feel the card being held in place. Then, remove your hand still keeping the tumbler held over the sink. Eventually the seal will be broken and the water will tumble out.

* The water is held in place by a combination of things. First and foremost the water is held inside the tumbler because the pressure of the air outside the glass against the cardboard was greater than the pressure of the water against the cardboard. Surface tension and cohesion also are involved. The card must be tightly sealed against the glass. This seal is accomplished by the surface tension and the cohesive properties of water.

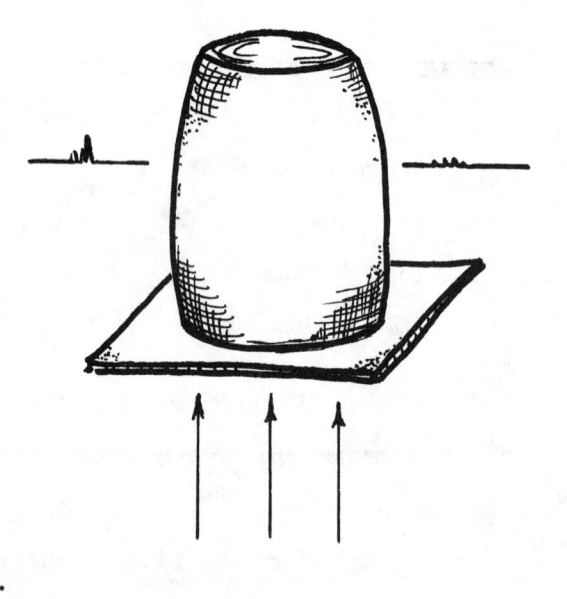

This activity can be further extended by repeating the exercise. Instead of using a solid card, using a thumbtack, punch several holes in the card. Have the participants infer the results.

Sound

Tie two pieces of light string to each end of an all-metal coat hanger. Bring the strings up to ear level, touching the ears with the string. Bend over slightly allowing the clothes hanger to hang freely. Have another person strike the hanger with a spoon or some other metal object. Vary the striking object.

* The sound generated by the spoon striking the metal hanger is carried up the string, ending in the ear.

Inertia

Place a coin in the center of a 3 x 5 card placed over the mouth of an empty tumbler. Quickly flick the card using your forefinger. The coin should drop into the tumbler.

* Newton's first law - A body at rest remains at rest (or in motion with uniform velocity) unless it is acted upon by an unbalanced force. This property of all real bodies is called inertia. The card-coin tumbler demonstration is a classic example of inertia.

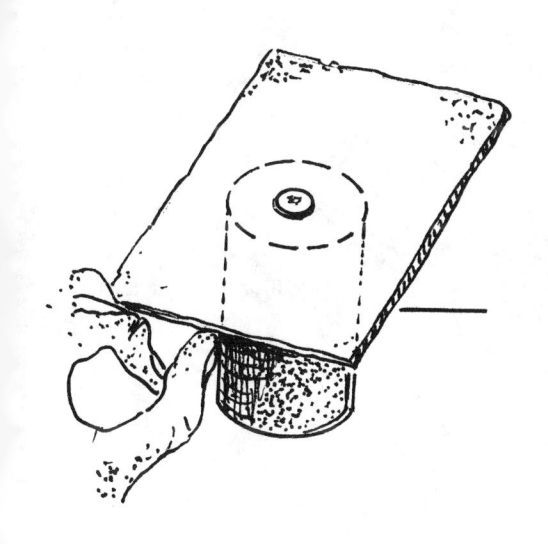

Surface Tension

Half fill a clean container with tap water. Sprinkle some talcum powder on the surface of the water. Dip a toothpick into a detergent solution and lightly touch the talcum powder. The talcum powder should split or rupture and move away from that point where the detergent is touched.

* Adding the detergent altered the surface tension of water. Talcum powder no longer is floating on a skin (surface tension) of water but on a skin of detergent water. This is a new condition. The floating talcum powder adjusted to this new situation which resulted in its rupturing.

Heat - Expansion - Contraction

From a piece of cellophane or thin plastic sheeting cut a small replica of a fish approximately two inches in length. Cut the body to show a broad and generous stomach. Place the piece of plastic flat on the palm of one student's hand. Wait and observe. Describe what you see..

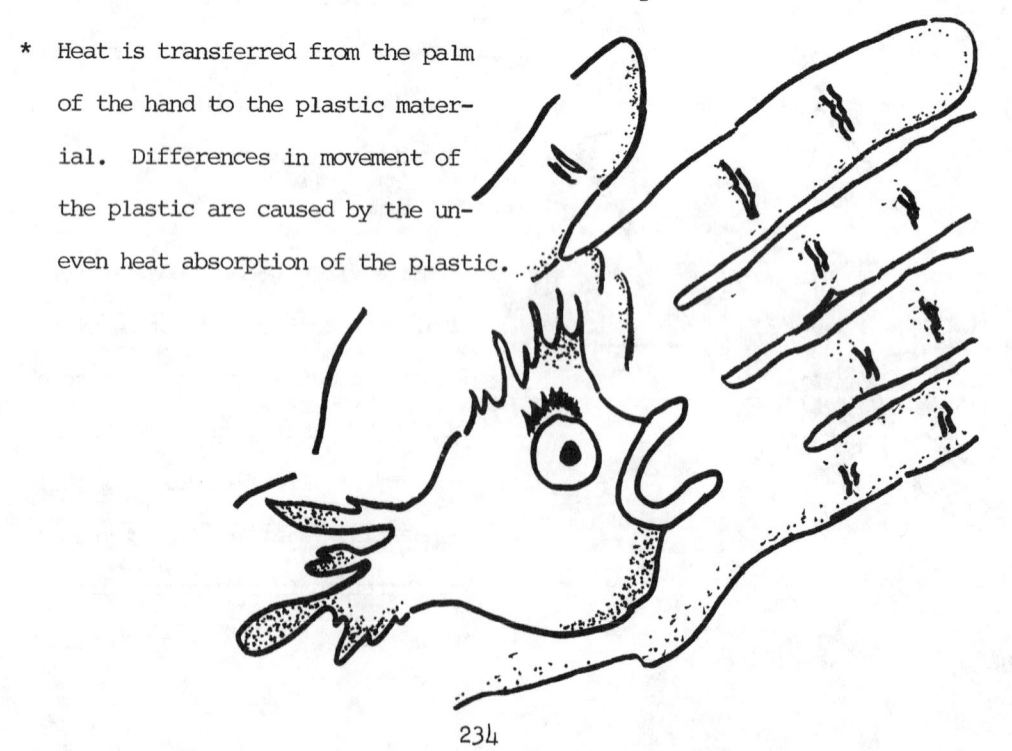

* Heat is transferred from the palm of the hand to the plastic material. Differences in movement of the plastic are caused by the uneven heat absorption of the plastic.

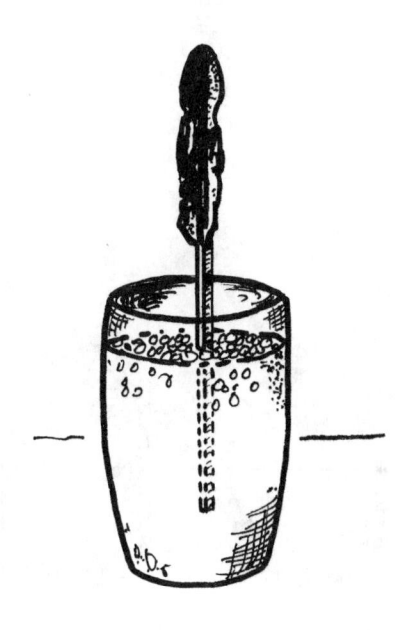

Friction

Fill a glass jar full of raw rice. Using a table knife, plunge the knife into the jar of rice several times. The drama increases as your action reveals little in terms of observable results. Plunge the knife in one more time and slowly lift. The whole jar will be lifted up.

* The rice becomes packed so tightly as a result of the previous attempts that it provides enough friction to lift the jar.

Water Molecules

Fill two, 200 ml beakers with water, one with cold water and the other with very hot water. Arrange these so that both can be observed simultaneously. Drop 4 drops of ink into each beaker. Observe the motion of the ink in both beakers. How do they compare?

* The ink will move most rapidly in the hot water where the water molecules are moving most rapidly as a reaction to the higher temperature of the water.

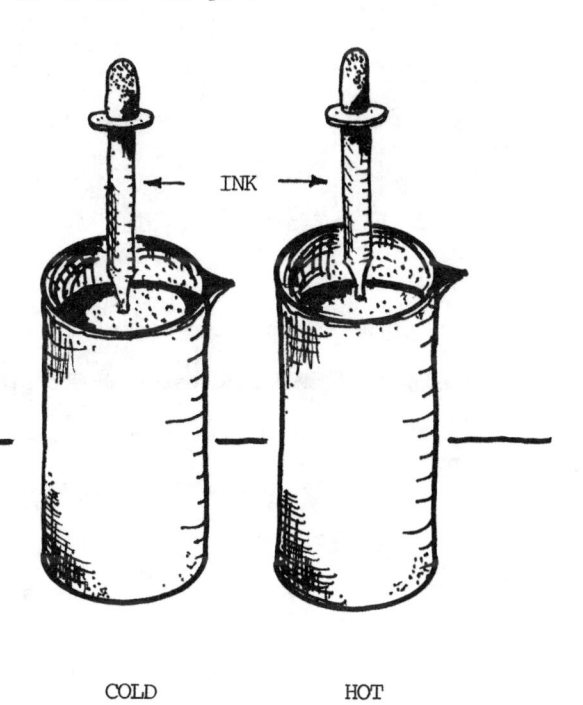

COLD   HOT

235

Density of Water

Obtain two bottles of equal size, shape, and volume. Fill one with cold water. Fill the other with hot, food-coloring, colored water. Do this in a catch basin or sink. Place a 3 x 5 card over the mouth of the cold water bottle. Invert the bottle holding the card tightly so no water runs out. Place the bottle and the card atop the warm water bottle. Make sure the mouths of the two similar bottles match up. Now slip the card out and hold the two bottles together at their mouths. What do you observe? Reverse the order, place the cold bottle on top. How does this affect the results?

* Cold water is more dense than hot water. The cold water should sink and the warm water should ascend.

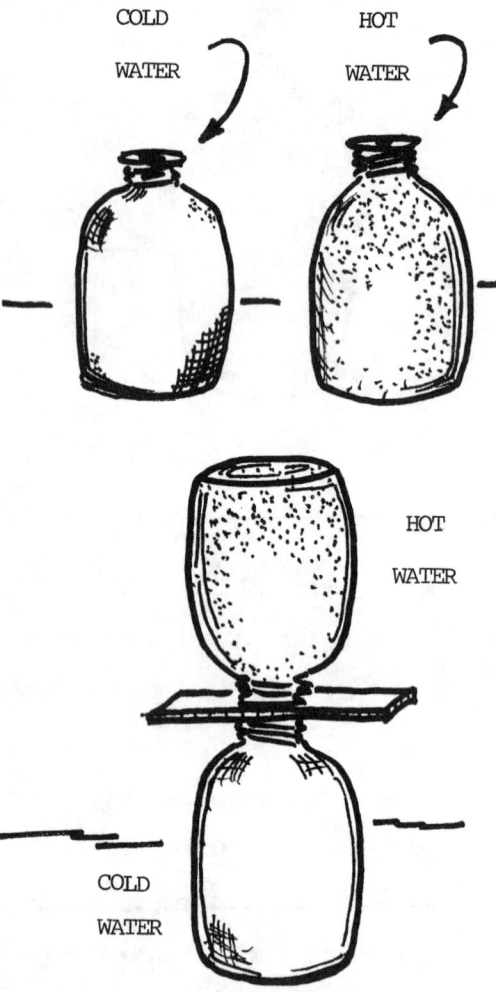

This same demonstration can be used substituting salt and fresh water for differences in temperature. Salt water because of the presence of dissolved solids is slightly more dense than fresh water.

Bouyancy - Density

Fill a glass with a clear, carbonated beverage. Quickly place four or five raisins in the liquid. The raisins will sink to the bottom of the glass. They will remain there for a short period of time, and then rise to the surface. There to perhaps gyrate a bit and sink to the bottom to repeat the cycle. What accounts for this observed motion?

* Carbonated drinks contain the gas, carbon dioxide. The raisins at the bottom of the container collect small bubbles of carbon dioxide gas. These bubbles buoy up the raisins...much like lifting a sunken vessel by pumping ping-pong balls into its hull. The buoyed raisins rise to the surface. Here, they lose some or all of the carbon dioxide bubbles to the air.

The raisin having lost its buoyancy is now too heavy to float and thus returns to the bottom where it collects more carbon dioxide bubbles. The raisin is again buoyed to the surface to repeat the cycle.

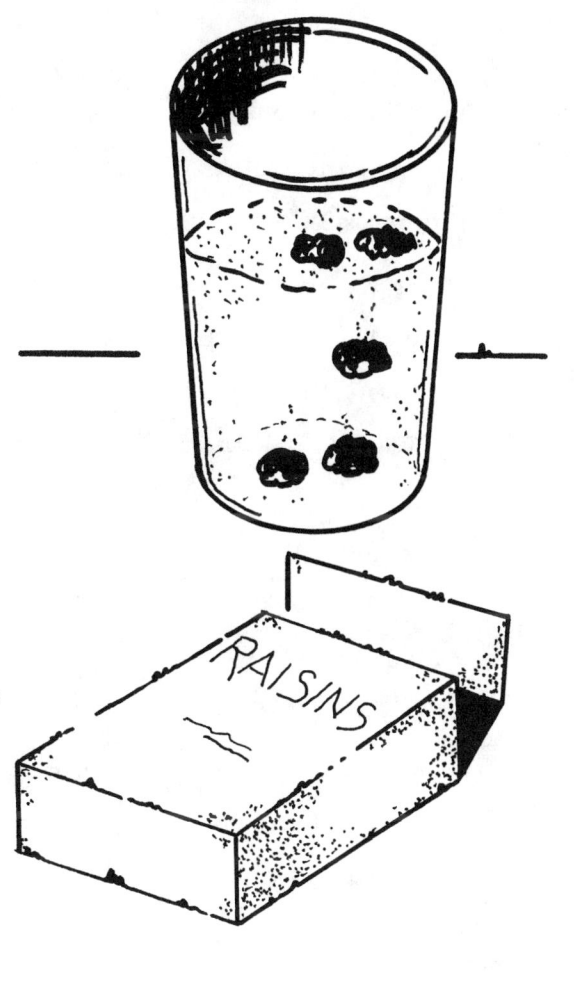

Air Pressure

Boil a round-bottomed pyrex Florence flask half full of water. Remove this from the flame and cork tightly. Using protective gloves, invert the flask and in this position slip the flask into the ring stand. Pour cold water on the flask. Be sure you have positioned a basin to catch the runoff water. The water inside the flask begins to boil without any additional heat applied. How can this be explained?

* As the steam is condensed by the cooling down of the flask by the cold water, the pressure on the water is lowered, and the water boils again.

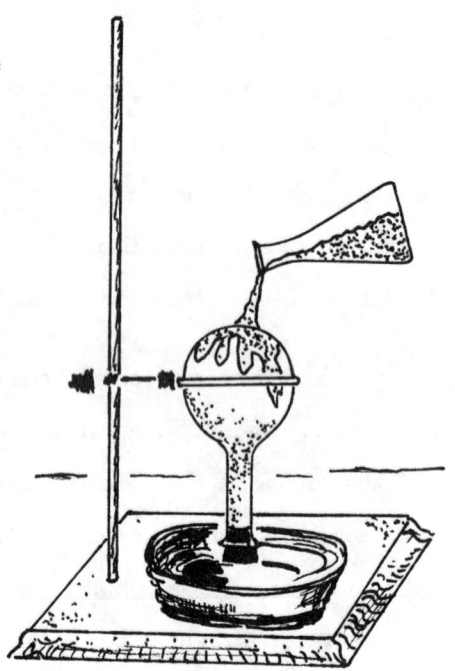

Static Electricity

Place a piece of glass across two separated books. Place pieces of paper underneath the glass. Rub the glass with silk or flannel material. What do you observe?

* A charge induced on the paper by the charged glass causes the paper to be moved about in an interesting way. When the paper loses its charge, it falls back down.

Heat - Kindling Point

Fill an unwaxed paper cup one-third full of water. Using a ring stand, clamp support, and wire screen, set the paper cup in position to be heated by a heat source such as a propane tank, candle or burner. Heat the water in the paper cup. The water will become warm, but the paper cup will not burn under these conditions. How come?

* Be careful! The paper cup will not burn as long as there is water in the cup. The paper does not burn because the water absorbs the heat and does not allow the paper to reach its kindling point, Once the water has dissipated, the cup will burn.

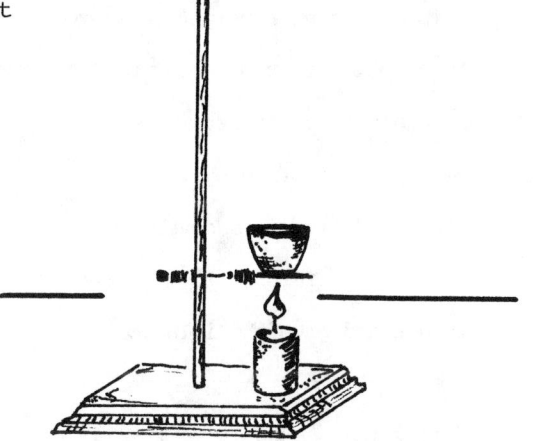

Matter

An interesting material can be fabricated from liquid starch (1/4 cup) and white soluble glue (1/2 cup) plus 1/4 teaspoon of table salt. If these proportions are too large use any one to two ratio (starch to glue) and a pinch of salt. This solid has some unusual properties which fascinate children.

1 PART LIQUID STARCH + SALT + 2 PARTS WHITE GLUE -----> = S^2G (ESQUARGEE)

Magnetic Attraction

You will need a ring stand, a paper clip, string, a magnet, a small cardboard box, and a weight. Arrange this as shown. The clip is attached to the string and positioned to fall within the sphere of the magnetic attraction of the magnet. Tension is kept on the string by pulling on the paper clip and using a weight to keep the paper clip from being pulled in against the magnet. The magnet, held in place by the ring stand and clamp, is covered by a small box. This is to hide the magnet and give the illusion that the clip is hanging vertically, independently in the air. Slight tugs on the string will cause the clip to bounce back to an upright position. If the distance pulled exceeds the magnetic field, the paper clips fall to the base.

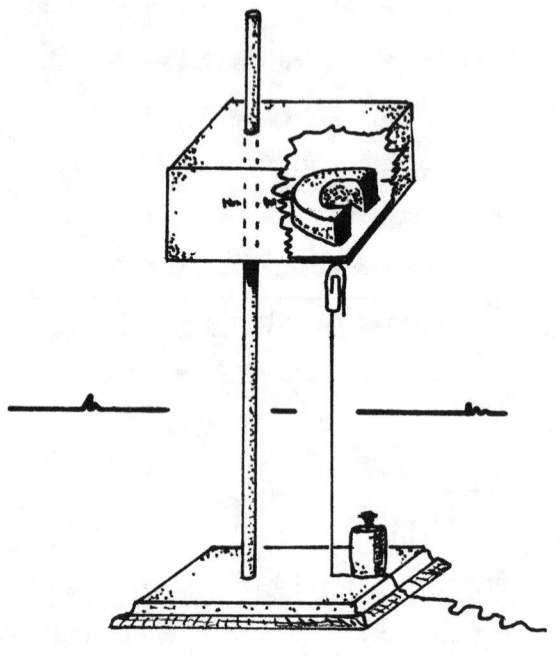

- Suspend one securely fastened bar magnet from a string. If this is permitted to swing freely, over time it will orient itself in space in a north/south direction.

Air Pressure - Compressed Air

You will need a potato, a pencil, and a soda straw. Place the potato on a flat hard surface such as a table top. Pierce the potato with the pencil. Being rigid, the pencil poses few questions as to whether or not this can be accomplished. Now, the question is "Can a soda straw be pushed through a potato much in the same manner as the pencil? This can be easily accomplished by capping one end of the straw tightly with your finger and pushing it into and through the potato.

* When the straw is capped, the air inside the straw is compressed making it rigid and strong.

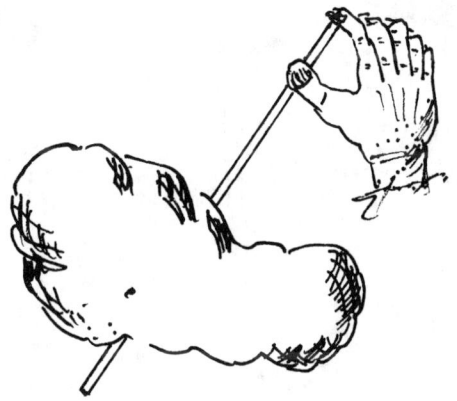

Water Molecules

Compare the two situations. Fill a water tumbler as full as you can get it without spilling the water over. If it does spill over, dry the exterior of the tumbler. One at a time, place as many cotton puffs as you can into the water. Keep count. When the water overflows, remove the cotton puffs. The final number of puffs represents how many puffs the water is capable of holding. Squeeze the total number of puffs into one solid lump. Fill another tumbler with water to the same level as previously used. Can the single lump of accumulated puffs, which represent the sum total able to go into solution previously, now fit into solution? Why not?

* Placing one puff in at a time allow the fibrous portions of the individual puffs to fill in the voids that exist between the water molecules. The large lump does not allow for this accomodation.

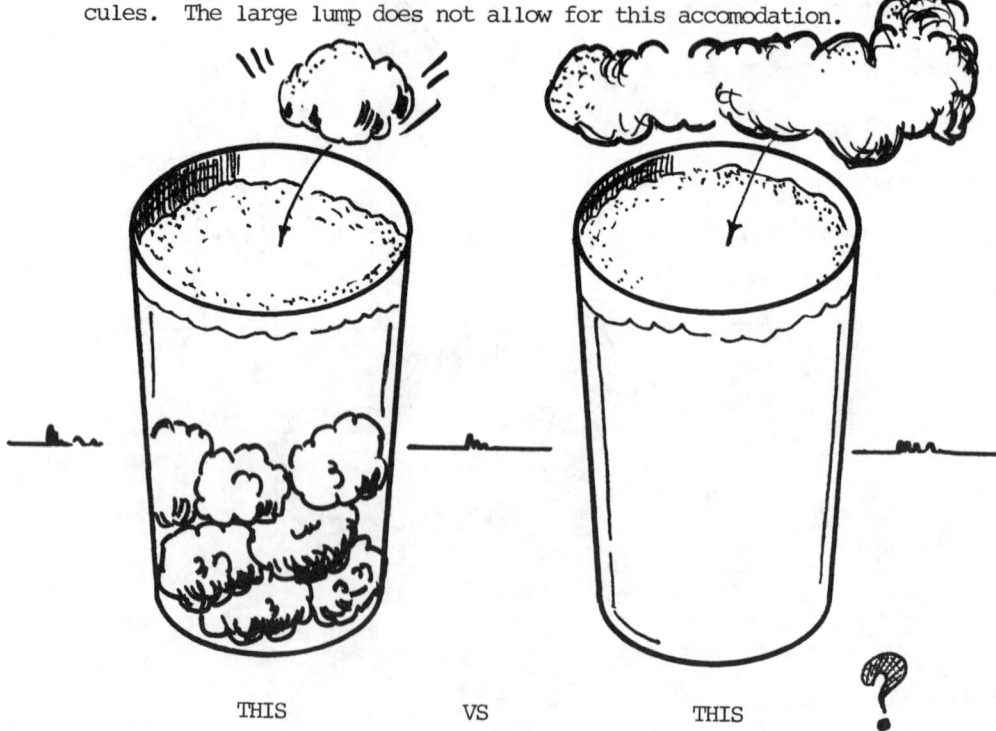

THIS   VS   THIS

Air Pressure

Obtain a one or two gallon tin can with stopper. Rinse it out thoroughly to make sure no residual volatile materials are contained within it. Put about a quarter inch of water in the can and heat the uncapped can over a hot plate or burner until the water boils and steam is seen exiting from the spout. Using protective gloves or cloth holders remove the can and quickly close the opening with the stopper making sure it is airtight. As the can cools, it crumples.

* Initially with the container open and water added, the air pressure inside and outside the can were equal. Heating the can forced most of the air out of the can causing a disequilibrium situation with air. When the can is capped, the air pressure inside is less than the air pressure outside the can; and the greater, outside pressure of air crushes the can as it cools.

Ocassionally someone has previously seen this time-tested demonstration and may appear disinterested. Add a new twist. Repeat the activity but prior to heating, drill an extremely small hole under or near the handle. The can will not collapse and this confounds the casual, blase observer.

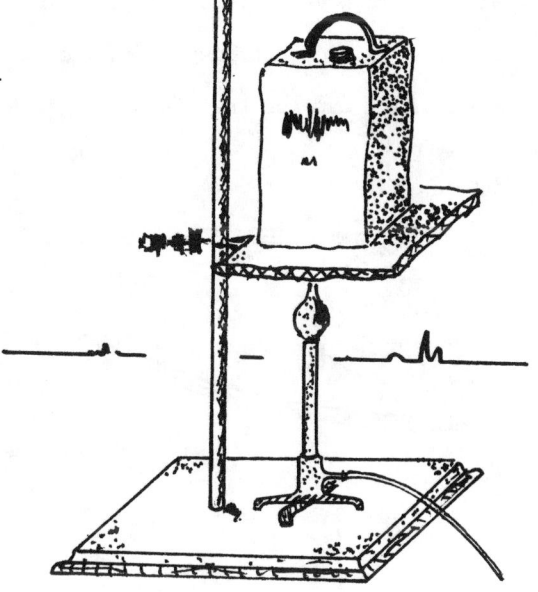

Differences in Expansion Rates of Metals

Obtain a bimetallic strip. This is two strips of different metals bonded together as one. The materials usually consist of iron and brass. When placed in a flame, the bonded strip (strips of iron and brass rivoted together are a suitable homemade substitute) will bend in one direction. Unbeknown to an observer, rotate the bimetallic strip after it has cooled. Reinsert this into the flame and the bimetallic strip bends in the same direction but by virtue of the rotation, it now appears to bend in an alternate direction. How can this be explained by the viewers?

* Using two metals bonded or joined together and inserted into a flame, the materials become hot and expand. Inasmuch as the strip is composed of two different metals, they expand at different rates. They are said to have different coefficients of expansion. The one material that expands the most is pulled in the direction of the other metal which expands the least.

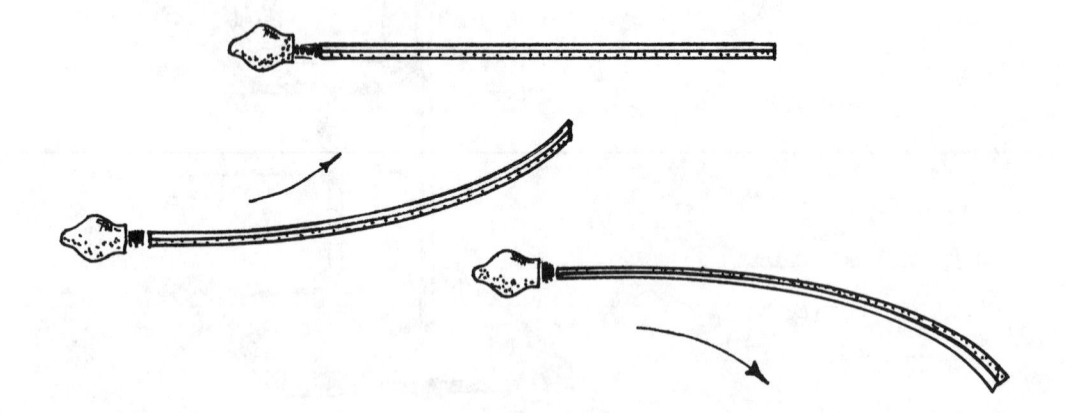

Air Pressure

Collect two small, glass soda-pop bottles. Keep one bottle at room temperature. Chill the other bottle by placing it in your refrigerator. Wet the rim of the bottle. Place a dime over the mouth of the room temperature bottle. Clasp your hands around the bottle. Observe. Repeat this using the chilled bottle. Compare your observations.

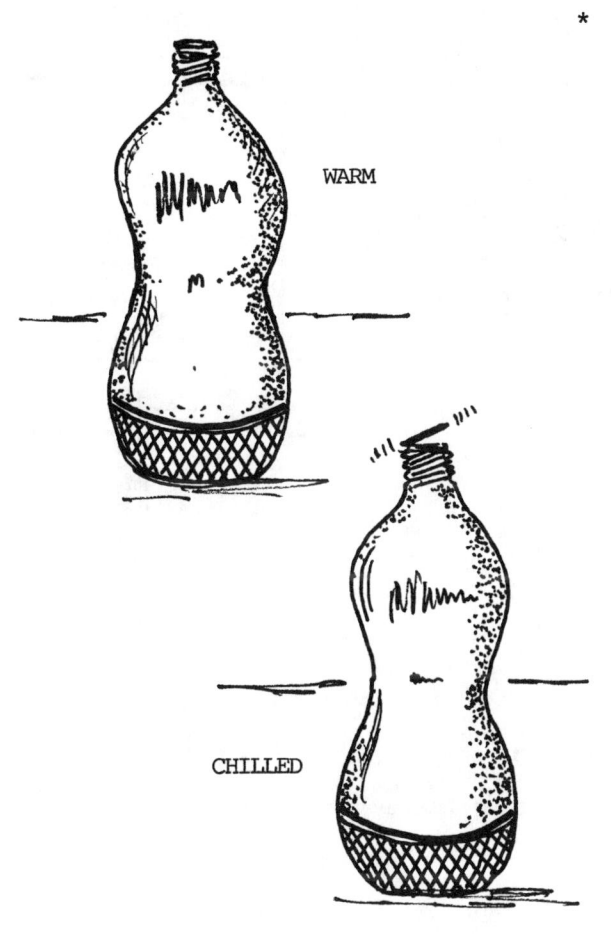

WARM

CHILLED

* Nothing will happen with the room temperature bottle. However, the chilled bottle's actions are more demonstrative. Warm hands placed around the chilled bottle will warm up the air inside the bottle. It expands. And, the dime will dance atop the bottle. A little water on the tip of your finger rubbed around the lip of the bottle prior to placing the dime on assures better results.

Optics

Draw a bird cage on a white card about two inches square. On the reverse side of the card draw a picture of your teacher and color your teacher with bright colors. Take a stiff soda straw and with scissors cut a slit on one end, down about one centimeter. Slip the card into the slit and thread a straight pin through the card, the soda straw, and the card again to hold the card firmly and erect. Place the soda straw between the palms of your hands and roll it quickly with a back and forth motion. What do you observe? How can you explain this?

* The rapid rolling motion of the straw is necessary to spin the card. An image will remain on the retina of the eye for only about 1/16th second after the object has been removed. This phenomenon is called persistence of vision. With the rapid rotation of the two views shown on the card, and persistence of vision, the teacher appears to be encaged.

Air Pressure

You will need a pyrex test tube tightly capped with a one-hole rubber stopper through which a piece of glass tubing, which has been cleared of any obstructions, has been inserted. The tubing should extend beyond the stopper inside the test tube about 1 cm. The external portion of the glass tubing needs to be approximately 10 cm in length. Place a small amount of water in the test tube. Bring the water inside the test tube to a rolling boil. Quickly invert the test tube. Stick the inverted test tube's glass tubing into a container of room-temperature, colored water.

* When the test tube is heated to where the water is boiling, the air pressure is reduced driving some of the air out of the test tube. In its inverted submerged position a partial vacuum is created. The air pressure on the surface of the water in the container is greater than the air pressure inside the test tube. An adjustment to establish equilibrium pushes the water up into the test tube.

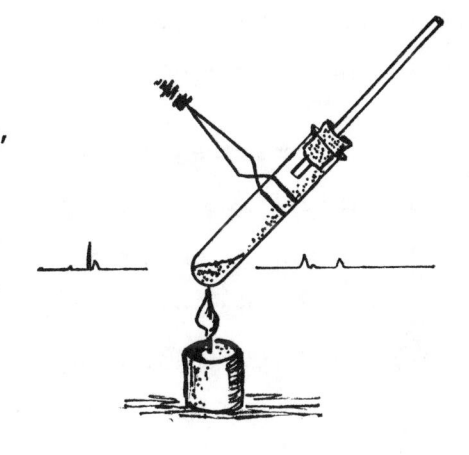

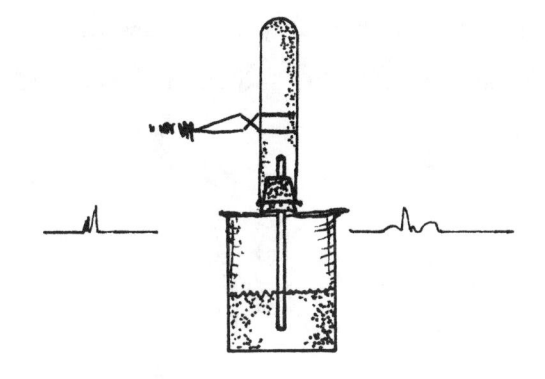

Air Pressure

Obtain a pyrex flask and a balloon. Place a small amount of water in the flask. Cap the flask with the balloon. Heat the flask and its contents over a hot plate or a propane burner. The heated water will expand and the balloon will be inflated. Set this aside to cool. The balloon, when cooled, will return to its original position and configuration.

Obtain a similar pyrex flask. Place a small amount of water in the flask. Using a flask holder and gloves, heat the contents of the flask. When the water comes to a boil, cap the flask with a balloon. When the balloon cools (placing it under cool water speeds up the process) the balloon shrinks, goes inside the flask and inflates itself inside out.

*In the first instance, the heated water expands the gas and it is forced up and inside the balloon. It is inflated. This is a closed system. When the heated gas cools, the balloon contracts and returns to its original state. In the second instance an imbalance is induced. The heated water heats the gas inside the flask, it expands and escapes the flask. Thus, the air pressure is reduced. Then by quickly capping the flask with the balloon we have created a situation where the air pressure external to the flask is greater than the air pressure inside the flask. Thus, the balloon in reaction to equilibrium is pushed into the flask and inflated.

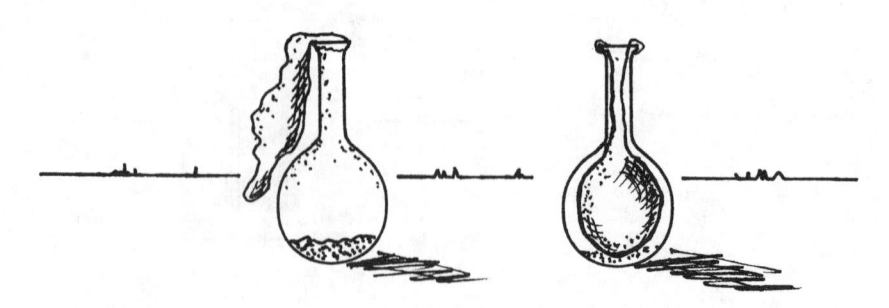

Air Currents

By means of a wire wrapped tightly about a candle, lower the candle down into a glass milk bottle. Compare your observations to the lowering of the same candle into the same bottle (aerated, of course, to clear out the bottle) but inserting a cutout block letter "T" into the bottle's opening.

* The cutout "T" separates the exiting warm air from the incoming cooler air. Without the "T" exiting warm air chokes off the chance for any cool, fresh air to come in and sustain the flame. For safety wrap the "T" in aluminum.

Color and Light

Mixing of colored lights can be observed by using various colored paints on buttons. Select as large a button as you can find. Paint radial segments alternately red and green. Thread string through two of the button holes. Set the button into a spinning motion by twisting and pulling the string in and out.

* The mixing of red and green lights reflected to the eye is yellow.

    Blue and red give purple.
    Green and blue give peacock blue.
    Red, green, and blue give white.

Air Pressure

Obtain a glass "T," some plastic tubing, and two balloons. Assemble as shown. Pinch off one balloon, exhale and blow up the other balloon. Pinch this inflated balloon off so that it retains the air inside it.

After unpinching the other tube, blow into the "T" deflecting the air into the uninflated balloon. Inflate this balloon to double the size of the first balloon. Pinch this off. Pinch off the mouthpiece leading to the "T." What will happen when air is allowed to flow unrestricted from one balloon to the other? Will things remain the same? Will the smaller balloon be inflated until the two balloons are equal? Or, will the air flow from the smaller balloon to the larger balloon making it even larger?

* The larger balloon gets larger. The smaller balloon has not lost as much elasticity as the larger balloon and when the air is free to flow in an unrestricted manner, the tautness of the smaller balloon causes it to constrict and push the air out and into the larger balloon whose elasticity is less. New balloons are more difficult than ones that have been blown up once or twice.

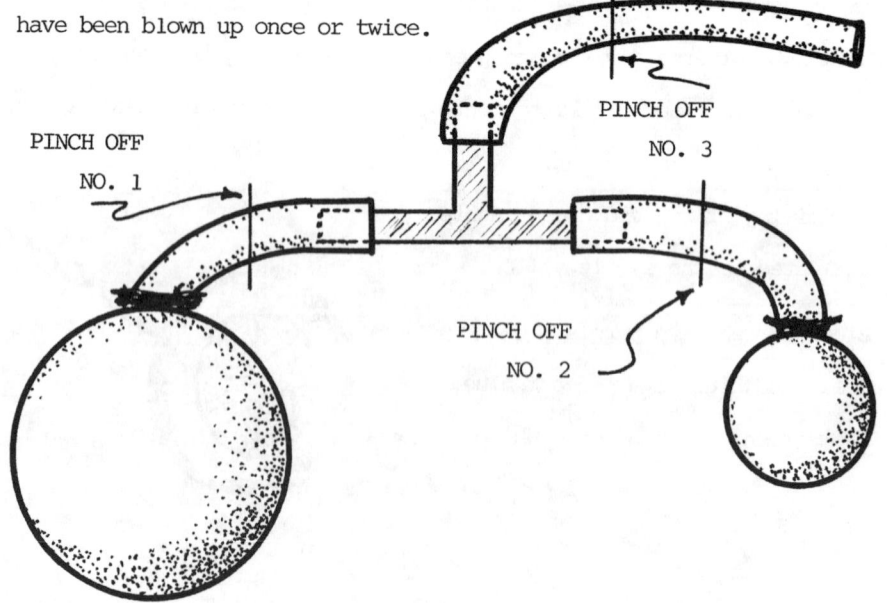

Air Pressure

Obtain a small hard-boiled egg that is slightly larger than the mouth of a glass milk bottle. Peel the egg and wet it; the water serves as a lubricant. Twist a 3" x 3" piece of paper into a rope-like shape. Light this with a match and while it is burning drop it into the bottle. Quickly place the peeled egg into the mouth of the milk bottle. At about the same time the burning paper goes out, the egg will slip through the neck of the bottle with a loud pop. How can this be explained?

* The burning paper uses up some of the oxygen from the air in the bottle and also drives some air out through expansion of the air through heating. With less air in the bottle the air pressure in the bottle is less than that outside the bottle. The egg quickly placed on the mouth of the bottle seals the bottle. The greater air pressure outside and behind the egg pushes it into the bottle. To get the egg out, the differences in air pressures must be reversed. This can be accomplished by tilting the bottle and rinsing out the burned paper with water. With the bottle in the tilted position, position the egg so that it rests in the neck of the bottle. Lean your head way back and press your mouth against the mouth of the bottle in an airtight manner. Blow as hard as you can. This forces air into the bottle. The air pressure inside the bottle now is greater than the pressure outside. The egg should start out slowly. Then be prepared to move quickly because it can eject itself with a strong velocity. Also, have a container positioned to catch the ejecting egg as it will fragment and be a chore to clean.

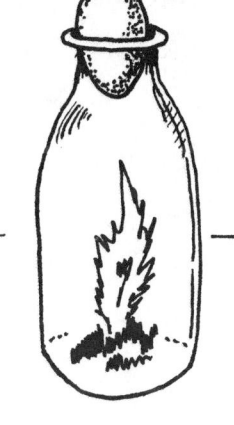

251

Air Pressure (Bernoulli's principle)

Obtain a plastic funnel. Place a ping-pong ball inside of it. Challenge the class to determine how high you can blow the ping-pong ball. With a little drama, huff and puff and attempt to blow the ball up and out of the funnel. The harder you blow, the more the ball is retained in the funnel. How can this be explained?

* This activity illustrates the principle of moving air streams, discovered by the scientist Bernoulli and called Bernoulli's principle. The fast-moving air streams create a low air pressure on one side of the ball. The greater air pressure on the other side of the ball then pushes toward the side with the lower air pressure. The top pressure on the ping-pong ball being greater pushes up against the ball holding it in place. This same activity can be repeated holding the funnel upside down and blowing through it. With a little assist from the palm of your hand until Bernoulli takes over, the ping-pong ball can be observed held up in the funnel as if by magic. The air rushing over the sides of the ball reduces the downward pressure, and atmospheric pressure then forces the ball up into the funnel where it remains as long as the air stream is maintained.

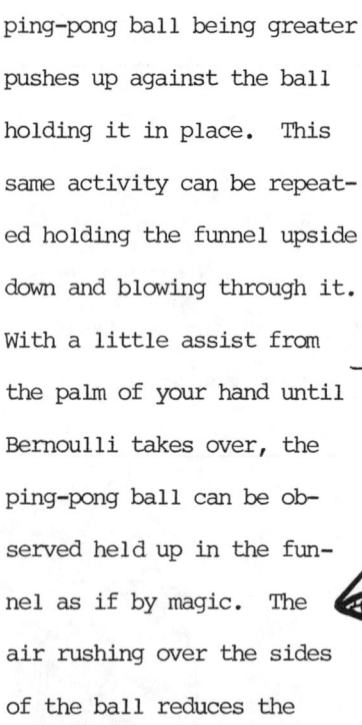

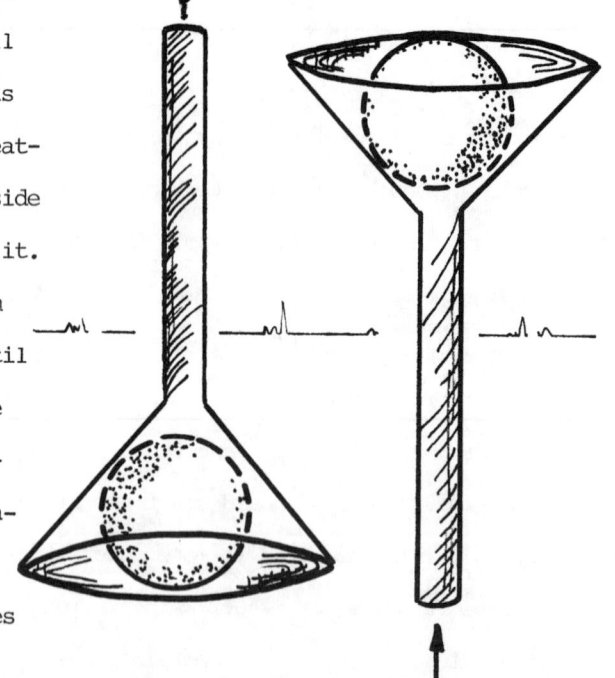

Effect of Air Pressure on Water

Obtain an empty, clear, plastic, liter-size soda container. Fill the plastic container, to the top, with tap water. Insert an eyedropper a little less than half filled with water into the container. If too much water is placed in the eye dropper, it will sink to the bottom of the container. The process would have to be repeated to be corrected. The water in the eyedropper will have to be reduced sufficiently enabling it to float. With the eyedropper floating at the top, cap the container. Squeeze the sidewalls of the container. What is observed?

* When the jar is squeezed, water is forced up into the eyedropper compressing the air inside the dropper. With more water contained inside the dropper, its density increases and it goes to the bottom. When the pressure is released, the compressed air inside the dropper is free to expand and it forces water out of the dropper. It becomes lighter and floats back to the surface of the water. This action can be observed by watching the fluctuating water level inside the dropper in response to external action on the container.

This discrepant event can be enhanced in several other ways. If children believe they understand the subsequent actions resulting from various manipulations, have them explain this. Instead of a dropper, cut the head off a small match (approximately 4 cm) about 4 to 5 mm in length. Place this small segment of a match into the container. Squeeze the container. The match segment should react much like the dropper. It should sink to the bottom and rise when pressure on the container is released. The same principle is involved. The structure of wood is cellular. Cells have air pockets which can be compressed. The wood takes on water and becomes more dense and sinks. Released pressure causes the compressed cells to expand, become lighter, and rise. It takes a bit of patience to get this to work. Sometimes the wood has to soak a bit before it becomes a functional diver. Nonetheless, it challenges students who feel they really understand the diving dropper to transfer that understanding to a new situation.

Try the diving dropper using a glass bottle instead of a plastic bottle.

Similar vessels; equal amounts of water; all capped by a rubber sheet; each having two test tubes (one smaller one inside a larger one) each arranged differently. Infer what will happen in these three situations when the covering rubber sheeting is pressed.

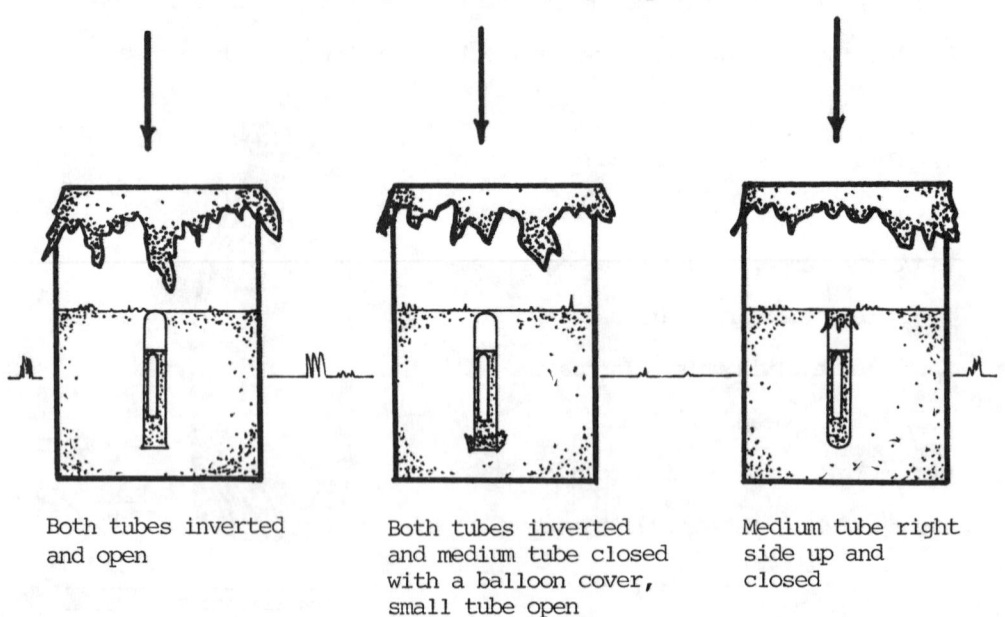

Both tubes inverted and open

Both tubes inverted and medium tube closed with a balloon cover, small tube open

Medium tube right side up and closed

Liquids (Cohesiveness)

Punch five holes close to the bottom and about five millimeters apart in a container. Place over a sink or catch basin. Pinch the jets of water together with your thumb and forefinger to form one stream instead of five. This single stream can be reformed into five separate streams by brushing your fingers across the holes in the can. Describe both situations. Explain your observations.

* There is sufficient electrostatic attraction to lace the five independent streams into one. The cohesiveness of water joins to maintain this unity.

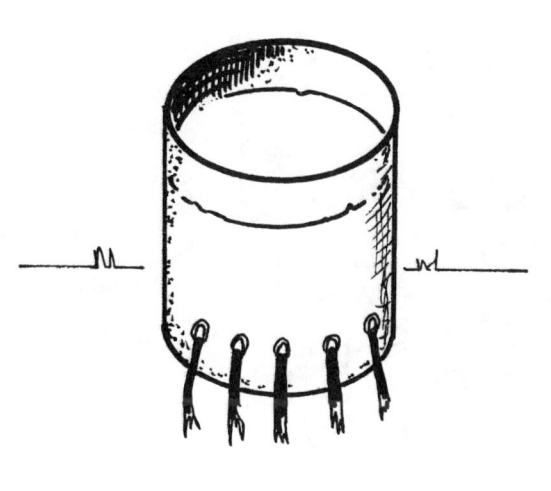

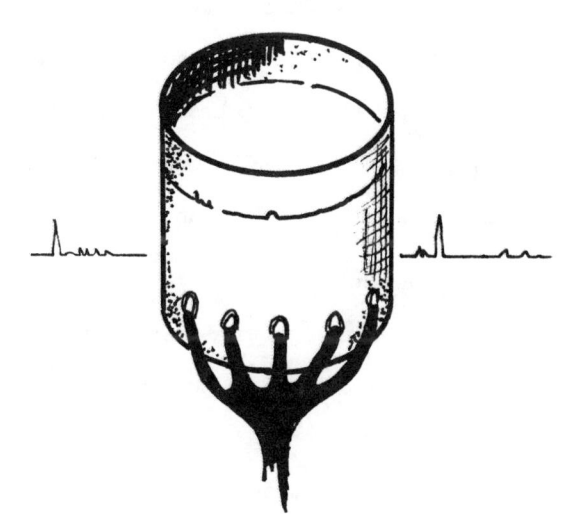

Air Pressure

Pour about two or three teaspoons of tap water into an empty cola can. With the top opening exposed to the atmosphere, heat the bottom of the cola can with a burning candle. Be careful to hold the cola can with tongs so as not to burn yourself. Heat the cola can until steam is viewed exiting from the top opening. Quickly invert the can and lower the open end of the can into a container of cool water below the water line. Observe what happens.

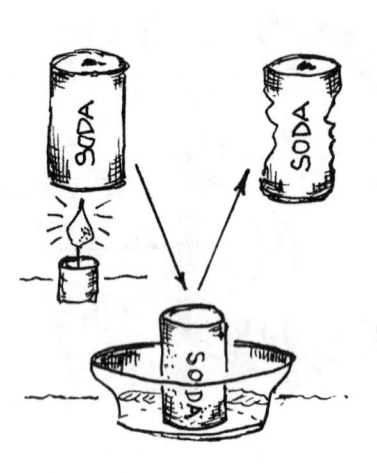

* Initially, the pressure inside and outside of the can was equal. Adding a small amount of water and heating the can caused much of the air to be forced out of the can. This reduced the air pressure within the can. Inverting and submerging the can in cool water sealed off the reduced interior, air pressure. Cooling action of the can, contraction of the air inside the can, and the existence of greater air pressure exterior to the can collapses the can.

Chemical Reaction

Place a raw egg in a glass. Add white vinegar until the egg is completely covered. Record your observations during the first few hours. One to three days later.

* Vinegar is a weak acid (acetic). An egg shell is approximately 95 percent calcium carbonate. These two ingredients react one with the other. Calcium carbonate will dissolve in weak acetic acid. Carbon dioxide bubbles will be released. Sometimes fresh white vinegar may need to be added to complete the process of shell removal. The remaining membrane holds the yolk and albumen in place. Describe the texture of these remains at the end of the third day.

PUZZLES AND PROBLEMS

- If you go to bed at eight o'clock in the evening and set the alarm to get up at nine in the morning, how many hours of sleep would you get?

    * One hour.

- Do they have a Fourth of July in England?

    * Yes, a fourth and a fifth, etc.

- Why can't a man living in Winston-Salem be buried west of the Mississippi River?   * He isn't dead yet.

- If a doctor gave you three pills and told you to take one every half hour, how long would they last?   * One hour

- A man builds a house with four sides to it (a rectangular structure), each side having a southern exposure; and a big bear wanders by. What color would the bear be?   * White, these conditions could only be met at the North Pole.

- A farmer had 17 sheep, all but nine died. How many did he have left?

    * Nine.

- Two men play checkers. They play five games and each man wins the same number of games. How do you figure that out?

    * They aren't playing each other.

- Take two apples from three apples. What do you have?   * Two apples.

- How many animals of each species did Moses take aboard the ark?

    * None, it was Noah, not Moses.

- A woman gave a beggar fifty cents. The woman is the beggar's sister, but the beggar is not the woman's brother. How come?

    * The beggar is a female.

- Two fathers and three sons left town. This diminished the population by three. How does one explain this?   * Each individual has several roles.

. A big Indian and a little Indian were sitting on a fence. If the little Indian was the son of the big Indian, but the big Indian was not the father of the little Indian what was the relation between the two?

    * The Mother.

Trace this illustration, drawing it with one continuous line, without intersecting the line or going over any part of it twice and without lifting your pencil off the paper.

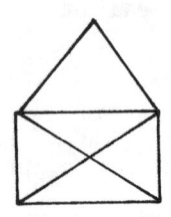

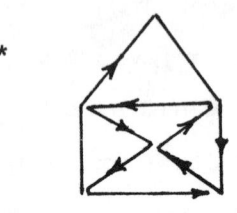

Trace this illustration, drawing it with one continuous line, without intersecting the line or going over any part of it twice and without lifting your pencil off the paper.

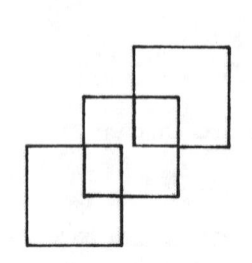

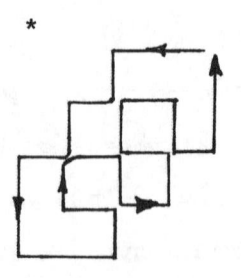

Join the sixteen dots with six straight lines, without taking the pencil from the paper and without going over the same line twice.

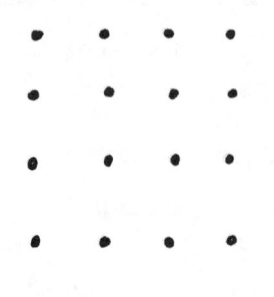

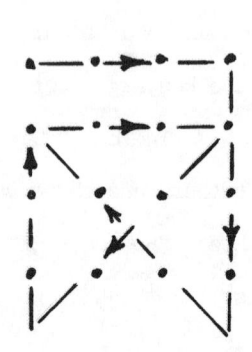

"Little Mikey appears to be growing twice as fast as any other child I ever knew," said his Father. "I'll say," said his Mother. "Two days ago he was three years old, and next year he will be six!" Explain this.

* The date of this conversation is January 1, 1983. One day earlier (December 30, was little Mikey's fourth birthday. Two days ago he was considered to be three years old. At the end of 1983, he will be five years old, and next year, at the end of 1984, he will be six.

A pen and a wooden pencil cost a total of $2.50. The pen cost $2.00 more than the wooden pencil. What is the cost of each?   * The pen cost $2.25 and the pencil cost $.25.

How much dirt is there in a hole 2 feet by 2 feet by 2 feet?
* None, the dirt is removed.

Lay a piece of rope or cord out straight on a table. Tie a knot in the rope without letting go of either end.

* Fold your arms first and then bend over and pick up the rope. When you unfold your arms, an overhand knot will appear.

Looking into a fenced-in yard where some people are petting dogs you are told that seven heads and 22 feet can be counted. How many dogs and how many people might there be behind the fence?   * 4 dogs; 3 people.

If it takes three minutes to boil an egg, how long will it take to boil seven eggs?   * Three minutes.

If you had twelve dollars and spent all but four dollars, how much would you have left?   * Four dollars.

Here are three playing cards that speak perfect English, one of them is a KING, who always tells the truth. Another is a QUEEN, who always tells the truth. Another is a JACK, who never tells the truth. No 1 says: "The King is next to me!" No. 2 says: "I am the Queen." No. 3 says: "The Jack is next to me." Name all three cards.

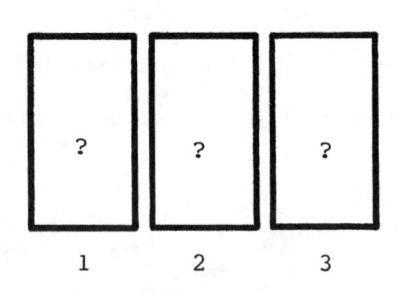

* The King never lies. Thus, he, cannot make either statement No 1 or No 2. The Jack never tells the truth, hence Statement 1 which is true is stated as a lie. By process of elimination, the Queen must be No. 3.

A farmer had three and 2/3rds haystacks in one field and six and 3/5ths haystacks in another field. He put them all together. How many does he now have?   * 1 large haystack.

Can you arrange for Jane to stand behind Tom and Tom to stand behind Jane at the same time?   * Yes, standing back to back.

A man and his two daughters must cross a stream. The man weighs 150 pounds and his daughters each weigh 75 pounds. There is a boat that can carry only 150 pounds. How will they cross the stream?   * Two daughters row over. One remains. The other rows back. The man rows over by himself. The one daughter rows back. The two daughters row back together to the other side.

There are two 5-gallon containers. One has 4 gallons of blue liquid and the other 4 gallons of red. One gallon of the blue is poured into the red container, and then one gallon of the now mixed 4-red to 1 blue is poured back into the blue container. Is there more red in the blue or more blue in the red container?   * Exactly the same amount of red in the blue as there is blue in the red.

- Something to think about:

    Why does a wet sponge soak up water more quickly than a dry sponge?

- An old trapper who was returning from the mountains with his pack animals heavily ladened with pelts was asked the price of his furs by a merchant he met along the way.

    "Well now," replied the trapper, "They are ten dollars apiece if you pick them out and five dollars apiece if I select them."

    What should the merchant's reply be in order to get the best deal?

    * RESPONSE: "I'll take the entire lot and you pick them out!"

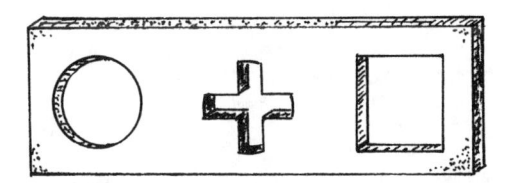

- All three cutouts from this block of wood are cut within an inscribed circle of the same diameter. Which of the three pieces will fit through more than one opening? What would a combination of any of these pieces look like which, if glued together, could pass through the original openings in the wood? The original outlines of the singular pieces must be preserved. and still be capable of passing through the openings.

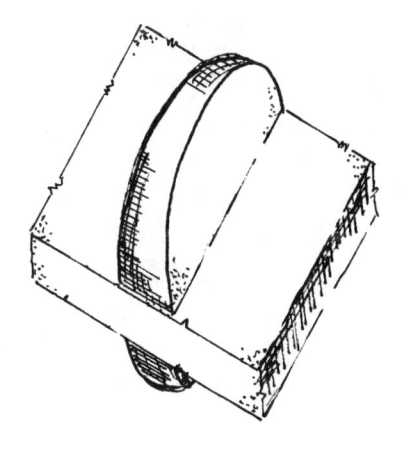

* All three of the singular-cut pieces should be capable of passing through any and all of the cutout patterns. A composite of the square and the circle glued in an interlocking manner will pass through all three cutout patterns. What other composites can be passed through the cutouts?

. A farmer opened his dairy barn one day and found that all but two of his milk measuring bottles were broken. All he had that were not broken consisted of a three quart bottle and a five quart bottle. A customer came to the barn with a two gallon container and asked for four quarts of milk. The farmer, using the two unbroken containers, had no difficulty measuring out the correct amount of milk. How did he do this?

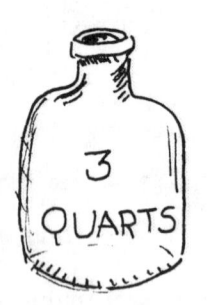

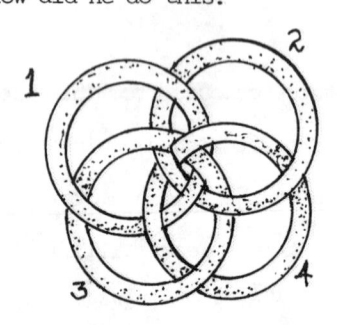

* Fill the five quart bottle, pour the contents from this container into the three quart bottle. This leaves two quarts in the five quart bottle. Pour this amount into the customer's two gallon container. Repeat this process one more time.

. Which of these rings must be cut to make all the rings come apart?

* Ring number three.

. A grocery man used an antique balance to sell his produce. He had a unique set of weights to weigh fruit as he sold it. These consisted of only five weights, yet he said he could sell fruit in bags containing any whole number of pounds of fruit up to 31 pounds. What were the numerical values of these weights? Which of these weights would he use to weigh out 13 pounds of apples?

* The numerical values of the five weights were 1 lb., 1 lb., 5 lbs., 5 lbs., and 20 lbs. He would have used 20 lbs on one pan, and 5 lbs, 1 lb, and 1 lb, plus the 13 lbs of apples on the other pan.

. Jones gave a hotel bellhop $15 for his cleaning bill. The hotel clerk found that Jones was overcharged and sent the bellhop to Jones' room with five $1 bills. The dishonest bellhop gave three dollars to Jones, keeping two dollars for himself. Jones has now paid $12. The bellhop had $2. This accounts for $14. Where is the missing dollar?

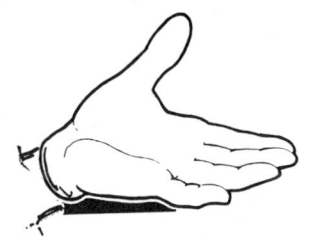

* Adding the bellhop's $2 to the $12 that that Jones paid produces a meaningless sum. Jones is out $12, which the clerk has $10 and the bellhop $2. Jones got back $3, which accounts for the full amount of $15.

. The diagrams show some equalities of weight among four different objects such as cubes, cylinders, cones, and spheres. One sketch is incomplete showing the balance in an unbalanced state. What is the fewest number of objects that can be placed in the right pan to balance the balance?

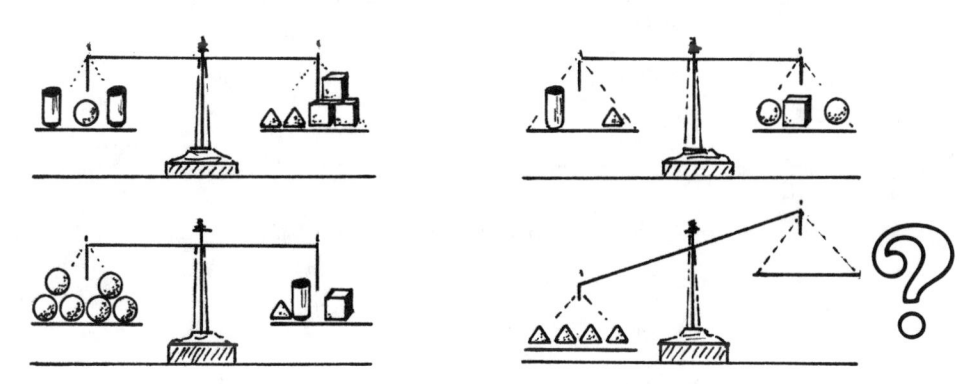

* One cube and one sphere. The relative weights of the solids are: cube 8, cylinder 13, cone 3, and the sphere 4.

Tenacious Think Abouts - Dilemma Science Activities
(Index)

- A Paleo-Indian Hunt   page 278
- What's Good for Whom?   280
- Who or What Belongs Where?   281
- The Missing Ring   283
- Engineering Noah's Ark   284
- Harriet's Holiday   287
- Bigfoot - Fact or Fiction?   288
- Where in Wheresville?   290
- How Much? How Many?   292
- How does one Build a Pyramid?   294
- Silent Guardians of Easter Island   302
- Vita Vista Revisted   307
- Centerport, U. S. A.   310
    - Another Fast Food Chain   312
    - The Best Garbage Truck Route   312
- Casino   314
- Where Depends on What?   315
- And Now What? No. 1   316
- And Now What? No. 2   317
- The Response should be...   318
- In Your Opinion   319

- Prescription for a Reef   320
- Stonehenge - Why and How?   321
- What Ever Happened to the Dinosaurs?   324
- Any UFOs Out There?   327
- Cause and Effect or
  "When is it my Turn?"   330
- Ancient Oriental Mariners or
  "Who Discovered America First?"   331
- Inane, Innocuous Inventions for
  Insatiable Inquiry   334
- From Kennebec to San Garbriel   336
- Sailor Beware   340
- The Delicate Dilemma   342

ENHANCING VERBAL INTERACTION BY LEARNERS THROUGH WRITTEN DILEMMA ANALYSIS

Demonstrations, student involved hands-on activities, and experiments comprise a substantial part of science instruction in the elementary schools. Demonstrations occur frequently in the primary grades. They occur less frequently in the intermediate grades. Hands-on activities, while occurring at all grade levels, are most prevalent in the middle and upper elementary grade levels. And, experimentation, if it occurs, appears to be reserved for the intermediate and secondary grades. A variety of reasons exist for this order. The concern is not with the hierarchical order of the structure, but rather with the paucity of opportunities that exist for critical thinking, problem resolution, and verbal interaction within this hierarchy. The inclusion of Dilemma Analysis in all grades can augment existing opportunities for the learner to be involved in verbal interaction and assist in the development of strategies for "thinking" one's way through a problem.

Verbal interaction between the teacher and the learner, and students interacting with the other students is brief or non-existing in many science programs. Regardless of the effort of the teacher as to planning, preparation, or the duration of the time alloted for instruction of a science topic, summary conclusions from children are usually sparse and terse. Well prepared vital lessons, punctuated with a great deal of teacher enthusiasm, often end on a flat note. Students' final summary response to such questions as "What do you think caused this? Does anyone have an explanation of that which was observed? Etc." usually comes out as one word, muffled contributions such as "Gravity." Any additional elaborations of such brief, verbal contributions are usually extracted by a continued, contrived dialogue generated by the teacher and directed to the respondent asking for clarification and an expansion of the initial contribution. Students' ability to analyze and verbalize science seem generally inadequate.

Written dilemmas are situational science activities that are an adjunct to on-going science programs. They are written, problematic situations which allow the participants to immerse themselves in a problem and to assess, organize, and interpret facts or data. Written dilemmas can take many forms. They can be situations formulated from an author's imagination. They can stem from revisions of current newspaper or magazine articles. Historical articles can also serve as a basis for written dilemmas. Written dilemmas can be short. They can be long. They can be simple. Or, they can be complex. Common to all is the fact that these problematic situations invite the participants to get involved and react. Written dilemmas are different than puzzles or situations that are convergent on one correct answer. By contrast, written dilemmas are open-ended, divergent, and invite verbal interaction as various interpretations are compared and discussed. Written dilemmas invite children to read, analyze, and to take an interactive position based on their interpretation of the available facts and data inherent in the dilemma situation. Each student can contribute because prerequisite knowledge relative to the dilemma is usually not required. Written dilemmas should be self contained. The basic ingredients necessary for subsequent interactive discussions should be contained within each dilemma. When leading evolving responses for specific dilemma situations, the teacher accepts all contributions. No penalty is attached to any explanation or stated position relative to the dilemma. The respondents are asked to provide a rationale for their thoughts. Thus, a free-wheeling, brainstorming style of science verbalization is constantly encouraged.

Dilemma usage depends on the grade level. Short span, high interest dilemmas with few constraints appeal to lower-grade level children. Longer span, broader interest-based dilemmas with multiple considerations appeal to upper-grade level students.

Dilemma Analysis involves the:

- identification of the problem.
- enumeration of the facts that have a bearing on the problem.
- formulation of strategies to uncover additional facts or statements.

ATTRIBUTES OF DILEMMA ANALYSIS

Dilemma Analysis stimulates thinking. Students are often accused of acting without thinking. Perhaps, this is because we expect them to think without acting or questioning.

Advantages that accrue from an inclusion of Dilemma Analysis in the elementary science curriculum are:

- Dilemma Analysis can be tailored for any audience. They can be short vignettes, or they can be extended into lengthy involvements.
- Dilemma Analysis enhances the skills of reading and listening. Dilemmas, in the early years, can be read to children. And, the verbal analysis can follow. When the skill of reading is acquired, children read dilemmas directly.
- Little scientific hardware is required to initiate Dilemma Analysis. Because of their ease of operation, Dilemma Analysis fit nicely into the day-to-day routine.

- Everyone can respond to dilemma situations. No pre-requisite knowledge is required to engage in the interaction.
- The dilemmas are purposefully selected and written for their open-endedness. Dilemmas invite a free-wheeling atmosphere for learning.
- The upper levels of Bloom's taxonomy (analysis, synthesis, and evaluation) are involved and promoted in arriving at inference and conclusions.
- Children enjoy the engaging situations. They are not intimidated by Dilemma Analysis inasmuch as there are no "right or wrong" answers only defensible responses to problems.
- Dilemma Analysis brings to light the fact that not every problem or situation is directly solvable, and that perhaps "more information" is needed. And, this search is projected through the consideration of appropriate questions such as, "If I had the answer or data about ... then ..."
- Dilemma Analysis can aid in the development of strategies for working through problems by delineating the problem, stating the facts available, making inferences derived from the facts and/or data, and asking oneself "based on what is given and in light of where I want to go with the problem, what would I like to have made available to me? What questions would I like to have the answers to? And, how can I obtain these answers?"
- Dilemma Analysis is efficient of time and money.
- Dilemma Analysis promotes teamwork, improves research skills as one expands beyond the limits of the original dilemma.

. It encourages children to evaluate optional explanations or solutions to the dilemmas.

The above statements reinforce those desired science instruction outcomes, grades K-6 as stated on page 1, ERIC/SMEAC Science Education Digest, Number 1, 1984, wherein the following desired outcomes were extracted.

. The integration of science with the teaching of reading and writing should be actively pursued.

. The ability to recognize problems, develop procedures for addressing the problem, recognizing, evaluating and applying solutions to problems.

. The ability to communicate orally and in writing.

Dilemma Analysis are brief confrontation with real or fictitious situations. Challenged to resolve the dilemma, children map out strategies, raise appropriate questions, and consider differing explanations or solutions for the dilemma.

Dilemma Analysis can be utilized by anyone. Dilemma situations can be prepared by anyone.

THE PREPARATION OF DILEMMA SITUATIONS

Topical items from newspapers and magazines are selected and rewritten in appropriate language and style for an intended grade level in accordance with the written dilemma format. A newspaper article can serve as the nucleus for a written dilemma. Prospective teachers read the article, winnow out that which is germane to the preparation of a dilemma situation, and then write an interactive dilemma for children.

*** A typical example explored, expanded, and rewritten as a Dilemma Situation for student analysis.

Water Diversion Splits States

Battle lines are being drawn between the Great Lakes states and the parched Southwest over water diversion.

While fresh water is taken for granted in the Great Lakes region, the Southwest wants more.

"It's going to be a hot political issue, and soon, if the population growth continues to explode in the Southwest," David Dempsey, environmental adviser to Michigan Gov. James J. Blanchard, said.

"What concerns me most is that the high water tables in the Great Lakes, coupled with demands for water in other parts of the country, gives you a recipe for a short-sighted diversion of water."

Debates about water rights are common in the West and Southwest. California and Arizona have fought about water in the Colorado River; South Dakota, Iowa, Missouri and Nebraska argue about the water from the Missouri River; New Mexico and Texas have been at odds on water rights to the Rio Grande and Pecos rivers.

Arizona is forcing residents to cut water usage below 1980 levels, and overall the West is the driest it has been in a decade. And demand for fresh water is escalating. Nationally, there was a 67 percent increase in water usage between 1950 and 1980.

Against that backdrop it is easy to understand national envy about the Great Lakes -- Huron, Ontario, Michigan, Erie and Superior. That system represents 95 percent of the country's fresh water supply and 20 percent of the world's surface fresh water. The Great Lakes contain more than 6 quadrillion gallons of water (that's a six followed by 15 zeros).

Discussions about diverting Great Lakes water to various water-poor areas have never reached a serious stage, largely because of the costs involved and widespread opposition from the eight states and two Canadian provinces in the Great Lakes Basin.

But that could soon change.

"I don't think diversion is a real possibility this year or next year. But the problem of water shortage, particularly in the Western states, is real and the pressure will not go away," William G. Milliken, the chairman of the Center for the Great Lakes in Chicago, said.

"In the next decade the issue of diversion of Great Lakes water will be very potent politically, economically and socially."

The reasons Michigan and other Great Lakes states oppose diversion are many. Along the shores of the lakes are power generating facilities which would have to go through costly refitting if water levels were lowered. The Great Lakes provide a major source of commerce; more than 225 million tons of cargo are shipped on them annually. If water levels were reduced by just one inch, it is estimated the largest ships would have to reduce their cargo by 200 tons.

>Indianapolis Star
>November 29, 1987

Some general considerations for developing a Dilemma situation for analysis by students:

- Read the article thoroughly.
- Define the problem or problems.
- Who are the involved parties?
- What are the facts?
- What are the relevancies of the facts?
- In some manner, prioritize the facts in terms of the problem.
- What groups does this problem impact on?
- Who gains? Who loses? Who gains what?
- What facts are missing?
- What other facts, data, etc. would be helpful to you in terms of proposed solutions to the problem?
- What are some suggested solutions?
- What, for whom, looks like the best possible solution?

With the above considerations uppermost in mind, the newspaper article needs to be re-written in an appropriate manner for the intended grade level of instruction. The finished product should be as nearly as possible self contained. This means sufficient information should be extracted from the original article (and in some instances embellished to make the involvement more challenging) to make it a worthwhile interactive engagement.

A SAMPLE DILEMMA SITUATION

Water, water, it is never where it oughter....be!

Water is a critical necessity of life. California and Arizona have fought about water in the Colorado River. South Dakota, Iowa, Missouri and Nebraska argue about the water from the Missouri River. New Mexico and Texas have been at odds on water rights to the Rio Grande and Pecos Rivers. People need water.

Some facts: Arizona is forcing residents to cut water usage below 1980 levels.

Nationally, there was a 67 percent increase in water usage between 1950 and 1980.

The Great Lakes contain more than 6 quadrillion gallons of water (that's a six followed by 15 zeros).

The Great Lakes contain 95 percent of the country's fresh water supply and 20 percent of the world's surface fresh water.

Eight states and two Canadian provinces border on the five lakes that make up the Great Lakes.

The population in the southwest portions of the United States continues to increase.

The water level of the Great Lakes appears to be rising (at least for the current period of time).

If the water level of the Great Lakes drops because of water being drawn off for the states of the southwest, numerous power generating stations would have to have extensive refitting to adjust to the new levels. Also if the water level drops just one inch, it is estimated the largest ships would have to reduce their cargo by 200 tons.

Questions: As you see it, what is the dilemma?

Is there an easy solution to the dilemma?

What are some major problems associated with:

> Leaving things as they are?
> Bringing water to the needed areas?

Who stands to gain from this action?

Who stands to lose from this action?

What would be some of the problems in engineering the transfer of needed water from the Great Lakes to needed areas of the southwest?

What solutions might you propose for this dilemma? What alternatives might you suggest? What resolution to the dilemma would be most satisfactory to a concerned majority of the people?

The above example is just that, an example. It could have been written taking many different routes, following many different styles and written for a different grade level. This is one route. One style. One level. Continued practice in selecting topics (from any source) and the preparation of dilemma situations will insure success.

Some additional examples are:

⬢   Jean and Joan are sisters. They received a set of play telephones for Christmas. They planned to install the phones between their bedrooms. The phones are battery-operated and these batteries must be inserted in a definite order or nothing happens. Each phone has a light bulb that lights up when the receiver is picked up. A bell rings when the opposite phone number is dialed.

The girls attempted to hook up the telephones. The lights went on for each phone when the receivers were picked up, but no bell rang at the opposite phone, and neither girl could hear the other.

Which of the following responses represents the best explanation of why the set did not work?

- The girls were not speaking loudly enough.
- The telephone operator was out to lunch.
- The batteries were dead.
- The batteries were inserted incorrectly.
- The wires from phone to phone were not properly connected.
- The girls were dialing the wrong number.

Do you have another explanation? Can you support your explanation?

⬢   You are sitting in the waiting room of your dentist's office waiting to have some dental work done. Your dentist has four small rooms as part of his office complex enabling him to work on several people at a time moving from one patient to the next. All doors to the four rooms are closed. While waiting you study the outer office and notice four seats had previously been occupied and the owners left the following articles in their individual chairs.

lunch box	brief-case	science magazine	novel
baseball cap	derby	cowboy hat	cowboy hat
1	2	3	4

You wonder about the owners. At that moment four doors open simultaneously and four men walk out. You study each one.

- One is elderly and dressed in old clothes all splattered with paint. His hands are rough and cracked. He smells of turpentine.

- One is middle aged and bald. He is well dressed. He is wearing a dark suit and a soft-colored tie. In general, he is well groomed. His hands are smooth and soft.

- One is young. He has an excellent tan on his face and arms. He has a crew cut. He is wearing western type clothing. He is wearing gym shoes.

- The remaining one is also young. His hands are rough and heavily tanned. He has on a short-sleeved shirt. You can tell from the difference between the tan below and above his wrist that he usually wears long-sleeved shirts when he is out in the sun. His face is tanned, but his forehead is not. He is wearing a college ring.

Can you match each man with the appropriate articles? Are there some matches more questionable than others? Which man do you think is a civil engineer? Why?

From these facts:

- I am over 18 years of age.
- I sometimes wear a gun.
- I sometimes wear a uniform.
- I protect my country.
- I love animals
- People never know when I am working.

Can you decide which of the following occupations best fits the given facts? A soldier, an FBI agent, a zoo keeper, a policeman, a mailman or an average citizen? Why is your choice a good one?

⬢ You are away at scout camp for the first time. You are deep in the woods. It is nighttime, and this is your first experience sleeping outdoors. You hear various noises but you can't see anything. You wonder what it is out there.

- You hear a scratching sound on bark. It seems to be coming from somewhere up in a tree.
- You hear clicking sounds similar to teeth snapping together.
- You hear a screech.
- You hear something running through the bushes.
- You hear what sounds like a struggle.
- You think it is either a fox or an owl. In the morning you discover a piece of brownish fur, a footprint that looks like it was made by a dog, and feathers.

What do you suppose happened during the night? Can you reconstruct the events? Did a dog eat a bird or a chicken? Or did a fox eat an owl? Or did an owl eat a fox? What additional information might help you explain these events?

⬢

A PALEO-INDIAN HUNT*

8,500 years ago a group of hunters on the Great Plains stampeded a herd of buffaloes into a ditch and butchered them.

Some Facts Are:

Buffaloes travel in large herds of 50 - 300 animals. Buffaloes have a keen sense of smell but poor vision.

A team of scientists had uncovered a ditch in western Nebraska which contained the remains of 200 buffalo. The ditch was approximately 200' long. Its narrow western end was only about a foot and a half in depth and about the same width, but it grew progressively deeper and wider to the east. Halfway down the ditch, its width was five feet and its depth six. Further on down, it was twelve feet wide and seven feet deep.

The ditch contained the remains of 200 buffalo. The bones were found in three distinct layers. The bottom layer contained thirteen complete skeletons facing south. These skeletons had their necks and front legs broken. Their front legs were drawn up under their chins. The second layer contained essentially complete skeletons from which a leg or two, some ribs, or the skull were missing. These bison had been partially butchered. The top layer contained single bones and parts of the remains. Sixteen calves were uncovered.

The surrounding area was composed of a reddish brown, crumbly rock.

The Paleo-Indians were very artistic.

During a feast a man could consume from 10-20 lbs. of dried meat.

About 75% of the buffaloes were completely butchered. On this basis, the total weight of buffalo meat would have been 45,300 lbs.

* From Alfred De Vito, "A Buffalo Stampede" Science and Children, Vol. 12, No. 2 (October, 1974), pg. 30. By Permission.

There was very little wood in the area.

The Plains Indian could completely butcher a buffalo in an hour.

Ten lbs. of meat per person a day were eaten. The remainder was preserved. The preserved meat and fat - together with the hides - would have weighed about 7,350 lbs.

Reconstruct the scene. Make a sketch of how you think the buffalo kill was planned. The direction the wind blew on the day of the kill. The direction of the hunters' drive.

Are all the facts provided relevant? Which ones are relevant? Why?

Infer:

- The month of the year the hunt took place.
- The manner in which the Indians butchered their kill.
- The choice of cuts of meat to be eaten on the spot.
- The probable number of hunters involved.
- How long they remained at this location.
- How much meat (weight) each member had to carry away.

WHAT'S GOOD FOR WHOM?*

Much wildlife has been killed in recent years as a result of man's technological developments. Many wild animals have fled into land not yet touched by man. Assume that in "Hidden Valley" in the Arctic Tundra, the last herd of "Milibar Moose" has found a safe home. Once abundant in the cold regions of North America, the animals have been driven away or killed by advancing civilization. This is their last stand.

The "Great Alabaster Oil Company" has discovered an oil deposit in Hidden Valley and is taking steps to begin drilling.

Permit the students to raise their own questions and concerns in this complicated situation through a role-playing situation. Some suggested questions you might pose to begin the debate are:

- Do we have a problem?
- Who will benefit from the oil deposit?
- Who will not benefit by it?
- What are some of the short and long-term gains?
- What are some of the short and long-term losses?
- How can we settle the problem?
- Do we need additional information?

With these and additional questions that you and the children raise, let each child or groups of children assume one of the following roles or a role of his own choosing. Attempt to resolve one or more of the suggested questions.

    Local government conservation official
    Head of "Save Our Wildlife" Society
    Driller for the Great Alabaster Oil Company
    Eskimo from nearby village
    Elected leader of the jurisdiction
    Townsman from town located near Hidden Valley
    Person in charge of discovering new oil deposits for the oil company
    Person from anti-pollution committee
    Secretary of Commerce for the area

* From Alfred De Vito, "What's Good for Whom?" Science and Children, Vol 10, No. 9 (January/February, 1973), p. 32. By Permission.

WHO OR WHAT BELONGS WHERE?

[Diagram of school site plan showing PARKING area on left, two OPEN GRASS rectangles in the center surrounded by parking spaces, shrubbery at the bottom, and a BASEBALL DIAMOND indicated at the upper right.]

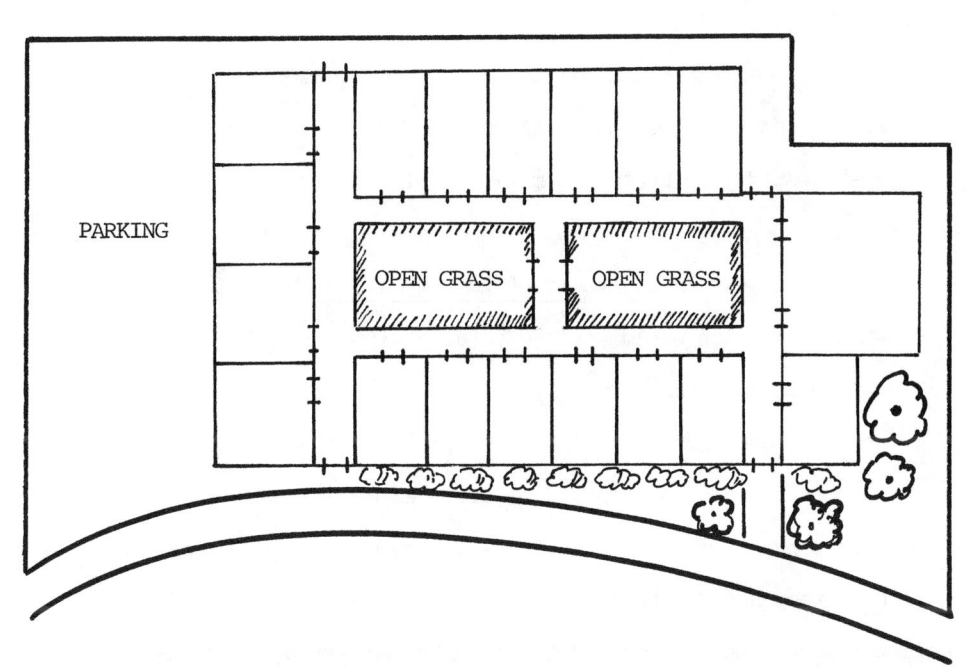

An unlabeled school building floor plan may be utilized as a situational science activity in a variety of ways. Have the children examine the school floor plan. You may wish to have your students label specific areas correlated with typical schoolroom assignments or you may wish to leave areas unlabeled. You provide as much or as little information necessary to involve your students at the level you deem suitable. As the arbitrator, you may assume one or more of the following roles: principal, secretary of the Board of Education, chairman of the Board of Education, architect, taxpayer, etc. Each role allows you to move the interaction forward or slow it down as needed.

One or more of the following questions (or better yet, conjure up

* From Alfred De Vito, "Who or What Belongs Where?" Science and Children, Vol. 12, No. 4 (January, 1975), p. 31. By Permission.

some new ones) may be asked of your students:

- What room assignments would you make for classrooms housing children in grades 1-6?
- Where should the principal's office be located?
- Where would you locate the nurse's office?
- Where would you locate the music room? The art room?
- Should the large area on the right be a gymnasium or a cafeteria?
- If the large room on the right is a gymnasium (or cafeteria) in which direction would you add on an additional room for a cafeteria (or gymnasium)?
- How many uses can you think of utilizing the open grass area?

The students' responses should be accompanied by a suitable rationale for their responses. The instructor, as the moderator, challenges or accepts each response depending on the selected instructor's "role" promoting thinking through an engagement with the suggested problems.

THE MISSING RING

A young girl had just received a ring for her birthday. The ring has a pearl set in it. She immediately put the ring on her right hand.

Later in the day she was invited to play tennis. At the end of the game she noticed that the pearl was missing from the ring.

If you were asked to help her search for the ring how would you proceed?

Some facts:

- The courts were constructed of concrete.
- It was early morning.
- It had rained the day before.
- Her opponent was a male.
- The girl was right handed.
- The courts have a slight depression (a small channel) all around it to assist in water drainage.

Utilizing these facts, how would you organize the search for the location of the pearl? Where is the most likely spot? The least likely?

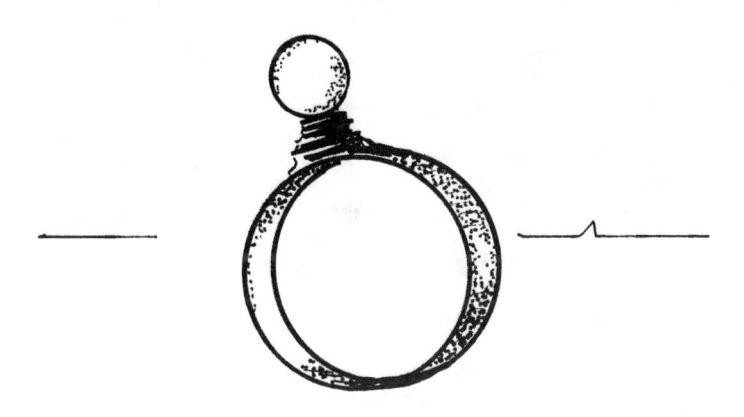

ENGINEERING NOAH'S ARK[*]

Widespread flooding was a common occurrence in the history of human existence on earth. Many geologists believe that around 4000 B.C. a flood covered what is now Iraq; archeologists have dug up cuneiform tablets in the region relating the tale of a man who survived by building a vessel. Could this have been the man called Noah?

As recorded in The Bible (Genesis 6:14-17), God commanded Noah to build an ark. The ark was to be built 300 cubits (150 meters) long, 50 cubits (25 meters) wide, and 30 cubits (15 meters) high. The ark was to have a door on its side, a skylight, and three decks, each deck to be divided into many rooms.

God warned Noah of a forthcoming flood upon the Earth. It was to rain upon the Earth forty days and forty nights. The flood would destroy every living thing. Noah was commanded to gather together at least one male and one female of each living creature (Genesis 6:17-20). These were to be placed on the ark along with Noah and Noah's family of sons and their wives. They all were to be kept safe on the ark until the flood subsided.

Suppose you were Noah and you agreed to design and build the ark, to assemble the animals of the Earth, and to feed and care for all these living creatures. How would you:

- Design an ark using the measurements and requirements provided. Do you think Noah's ark looked like a typical boat? What would your ark look like? Make a sketch or build a model of your ark.

    It is thought that the capacity of the ark was approximately 5,500 tons; enough for at least 1,000 pairs of animals. It is also

[*] From Alfred De Vito, "Engineering Noah's Ark," Science and Children, Vol. 11, No. 5 (January/February, 1974), p. 32-33. By Permission.

believed that Noah's ark had no sails, no oars, no engine, and no rudder. Thus, it could not be steered.

Meir Ben-Uri, one of Israel's leading religious painters and synagogue designers, reasons that the ark was probably built of either bamboo or lightweight wood, both of which are common to southern Iraq. Meir Ben-Uri reconstructed the ark relying entirely on God's commands to Noah. Ben-Uri concluded that the ark, in order to be built of such dimensions, using available materials, would have to be built in the shape of a prismatic rhomboid. This is essentially a long bar whose cross section is shaped like a diamond. The dimensions of the ark would make it larger than a World War II Liberty Ship.

- Noah's ark was to have three decks. Which animals would you place on the lowest level, the middle level, and the top level? What reasons can you give for placing the animals where you did?
- An ark 30 cubits (15 meters) high is a tall ark. How would you get the animals of the Earth up and into the ark? Once you got them up there, how would you get them to the level you wanted them inside the ark? Alphabetically, the aardvark would be first. But how would you organize two of every animal to go aboard the ark?
- In your design of the ark, you must provide for the storage of food for the animals as well as the humans on the ark. How will the appetites of the various living creatures of the Earth affect their distance from the food storage area? How much food will be needed? What varieties of food?
- Where on Noah's ark would be the best place for you and your family to be?

- "Praying" mantes feed on smaller insects. How would you protect certain insects from being eaten by other insects?
- How much waste would all the animals on Noah's ark generate? What would your pollution solution be?
- What are some other questions you would need to answer before putting the animals on the ark?

Some 4 to 5,000 years later (in 1955) Fernand Navarra, a French businessman and an explorer, found on Mt. Ararat in eastern Turkey an estimated 50 tons of wood buried in a glacial ice pack. Samples of this wood have been identified as hand-tooled oak, treated with a pitch-like substance. It is a mystery how this much wood got to this spot. The wood is located at an elevation of 14,000 feet. This is several thousand feet above the tree line and over 480 kilometers from the nearest trees of any size. Airplane pilots flying over Mt. Ararat have repeatedly reported a ship-like shadow in the ice. It is believed by Mr. Navarra that this wood is the original material of the ark Noah, mentioned in The Bible. What do you think?

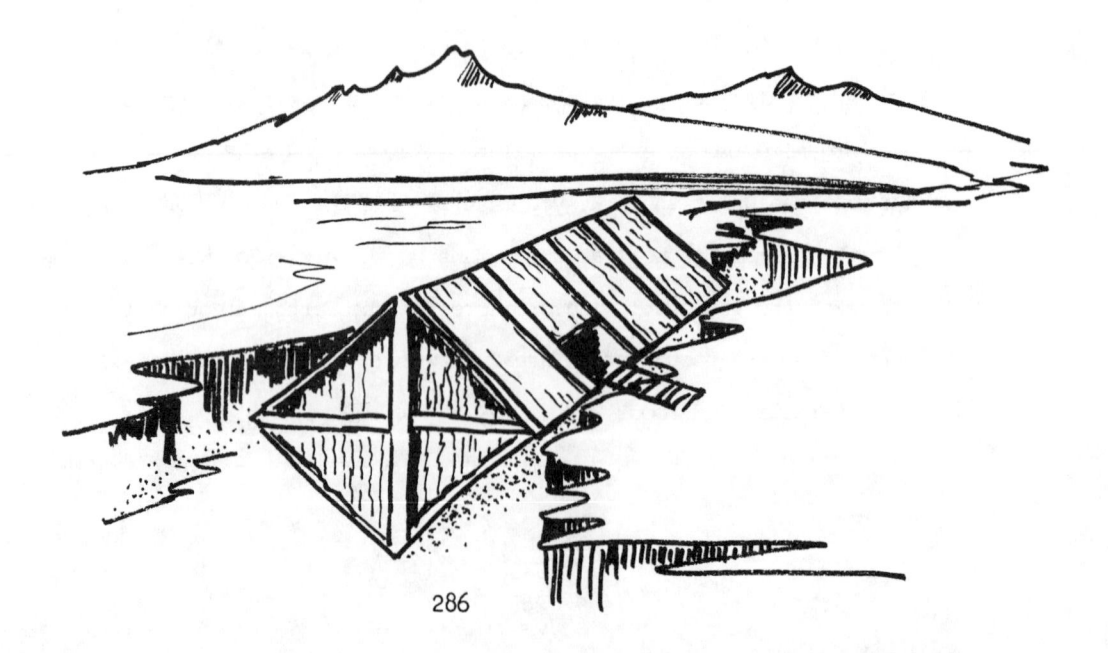

HARRIET'S HOLIDAY*

Nila and Donna are two sisters who lived in an apartment dwelling. One of their neighbors was a retired school teacher named Harriet. One of their other neighbors was a young couple whose parents wanted to visit them for a month. The parents of the young couple lived in Hawaii. The young couple looked for an apartment for their parents but could not find one for this length of time. They happened to mention this to the retired school teacher. The retired school teacher suggested to the young couple that she and the visiting parents swap apartments. She would live in their home in Hawaii for one month and they could live in her apartment for one month. The arrangements were made. Harriet departed for Hawaii. The next day the young couple's parents arrived. They lived in Harriet's apartment and drove Harriet's car. They stayed exactly one month. Nila and Donna expected Harriet home the next day. For two weeks they waited and still Harriet did not return or at least it appeared that she did not return.

Facts:

- Harriet's car remained in front of her apartment.
- Harriet's daughter lived in the same town.
- Occasionally the daughter was seen entering the apartment and occasionally she would start the car and allow it to run.
- It was football season. Harriet was a real football fan. She always attended the college games in town. If she had returned on time, she could have attended two of the remaining games.
- The town was a midwest town. Harriet had no other relatives than her daughter who lived in town.

Nila and Donna began to wonder what had happened to Harriet. What possibilities might account for Harriet not returning?

* From Alfred De Vito and Gerald H. Krockover, <u>Creative Sciencing: Ideas and Activities for Teachers and Children</u>, 2nd ed., p. 87, Copyright 1980 by Alfred De Vito and Gerald H. Krockover. Reprinted by permission of Little, Brown and Company.

BIGFOOT - FACT OR FICTION?[*]

Since the early 1800's, hundreds of stories have been told about giant, Earth creatures who look part human and part ape. These stories or legends spring from observations made by many people who claim to have seen such creatures. These creatures have been sighted in areas of the Himalayan Mountains in India where they are called "Yeti." They have also been sighted in the northwest United States and Canada. Here, these creatures are called "Sasquatch" or "Bigfoot." Are these creatures real? Or, have they just been imagined?

Some facts:

Many people claim to have seen Bigfoot. Bigfoot has been described as standing between six and eleven feet tall. Most people who believe they have seen Bigfoot describe him as being human-like, broad of shoulders, short necked, and hunched over. Observers further state that Bigfoot walks in an upright position much like a man and who, when observed, stops, stares, and appears not to be too aggressive. Bigfoot is said to have a cone-shaped head with a face that is monkey or ape shaped. It has been further noted that Bigfoot is covered with reddish-brown hair. While some photographs have been taken of Bigfoot, all the photographs are fuzzy and leave much to the imagination.

No Bigfoot skulls, skeletons, or waste droppings have ever been found.

The best evidence to support the possible existence of Bigfoot seems to be in the footsteps Bigfoot leaves behind. Over 300 Bigfoot footprints have been reported in the United States. Bigfoot's prints range from twelve to twenty-two inches in length and seven to eight inches in width.

[*] From Alfred De Vito, "Bigfoot - Fact or Fiction," Science and Children, Vol. 14, No. 7 (April, 1977), p. 25. By Permission.

An estimate of the size of such a creature necessary to leave such prints would be that this creature must be at least eight feet tall and weigh over 800 pounds.

Observations of Bigfoot have been made in every month of the year.

Bigfoot has been observed more often at night than during the day.

Early Bigfoot sightings (1924), as well as numerous recent sightings, have been confined in the United States to the Pacific Northwest territory - particularly in the Cascade Mountain Range. The area in which the most numerous sightings of Bigfoot have taken place is a 40 square mile region heavily covered with dense cone-bearing trees (spruce, pine, etc.). This area, which is entirely wild, includes parts of Washington, Oregon, and northern California.

Some Questions:

- What does Bigfoot eat?
- Why haven't we been able to find any traces other than footprints?
- Why haven't we been able to capture Bigfoot?
- Is Bigfoot an animal? Man? Or both?
- If we capture Bigfoot, what should be done with him?

Bigfoot or not - certainly something exists that needs to be explained. Could a creature this big be wandering through the forest in northwest America and still be unknown to science? What do you think?

WHERE IN WHERESVILLE?*

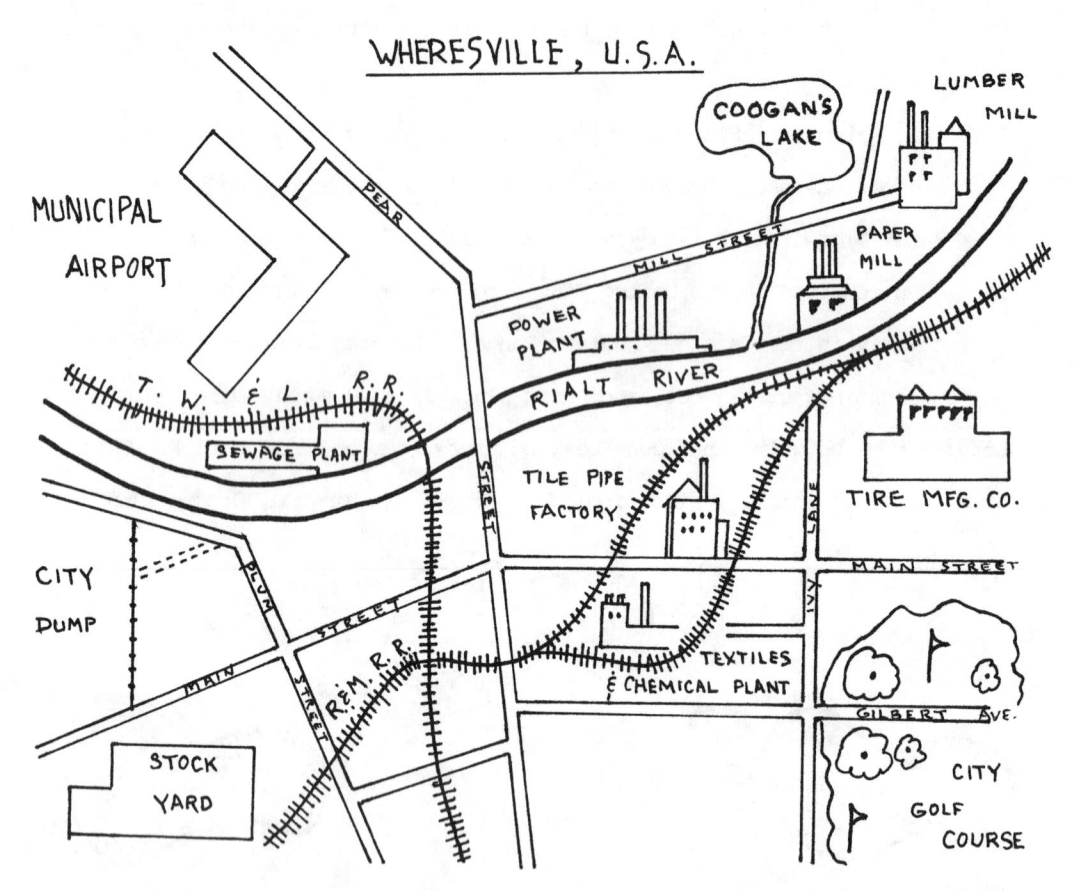

map of Wheresville, U.S.A. showing Municipal Airport, Coogan's Lake, Lumber Mill, Paper Mill, Power Plant, Sewage Plant, Tile Pipe Factory, Tire Mfg. Co., Textiles & Chemical Plant, City Dump, Stock Yard, City Golf Course, Rialt River, T.W.&L. R.R., R.&M. R.R., and streets including Pear, Mill, Plum, Main, Lane, Gilbert Ave.

Established early in the Nineteenth Century at the head of navigation on the Rialt River, Wheresville has become one of the old-new cities of the southeast. The city has large industrial operations in the manufacture of fiber and cast iron, soil pipe and fittings, paper from southern pine, rubber tires for motor vehicles, lumber, textiles, and chemicals.

Some facts:

. For clarity, only the major streets are drawn in. Numerous

* From Alfred De Vito, "Where in Wheresville?" Science and Children, Vol. 10, No. 6 (March, 1973), P. 36. By Permission.

streets crisscross between all these major roadways.

- The area between Plum Street and Pear Street is the downtown business center.

- Over 100,000 people live in this town. Eighty percent of the population live south of the Rialt River. Twenty percent of the population live north of the river.

Some questions to ask:

- If you were planning to move to Wheresville, where in the town would you like to live? Why? What factors are critical in deciding where one might choose to live?

- A student assumes the role of a businessman. He wants to build an amusement park. Where is the most likely spot for this enterprise? Why?

- The town wants to establish a bus system. They decide they need three, bus routes. Which are the most appropriate routes? Why?

- Where do you think the most valuable land in Wheresville is located? The cheapest? Why?

- Does the town need a park? Cemetery? Where? Why?

- The town wants to build two new elementary schools, where should they build them? (As the instructor you locate the appropriate position for the junior and senior high schools; you may also wish to establish the locations of the four existing elementary schools.)

- Suppose the town fathers do not like the present location of Wheresville; they want to move it ten miles upstream; how would you plan an entirely new town?

HOW MUCH? HOW MANY?*

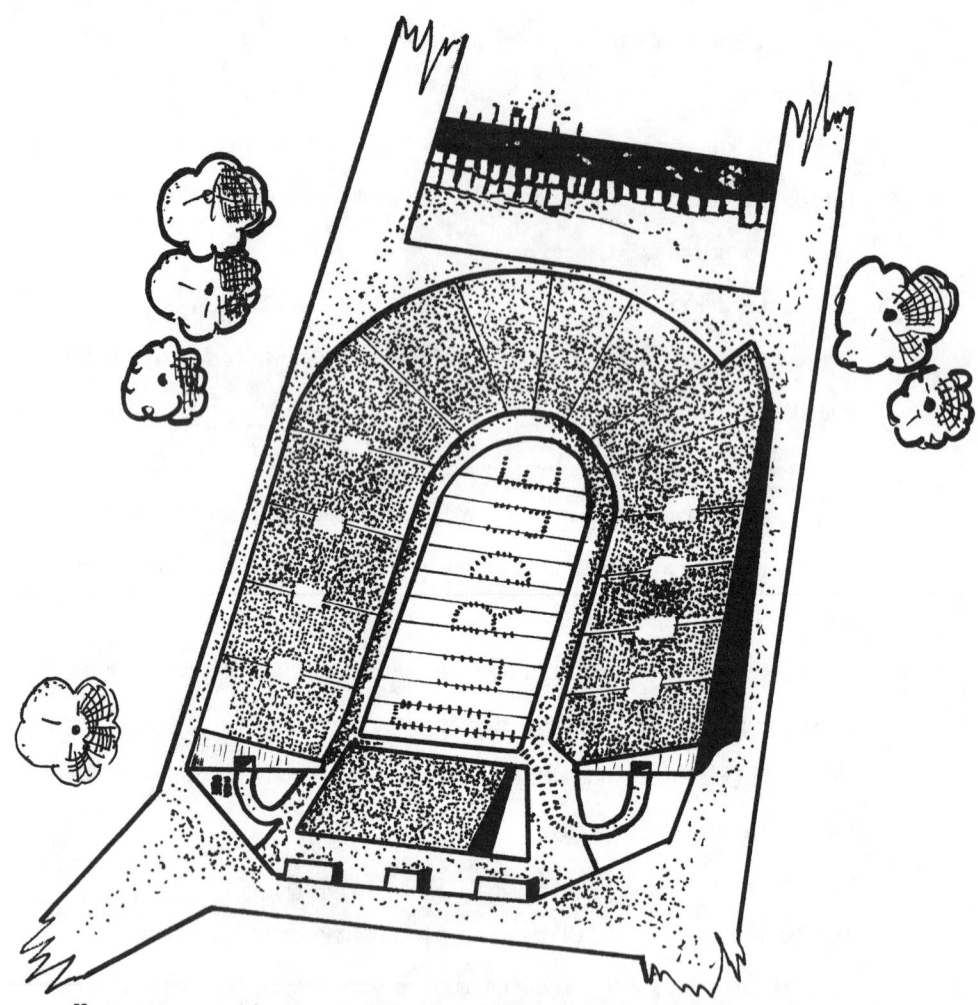

How many questions related to science and math can you generate by studying this drawing? How many can you answer? A few might be:

- Is the game about to start? Finish? Or is it half-time? How might you support your statement?
- There are about 66,000 people in attendance at this football

---

* From Alfred De Vito, "How Much? How Many?" <u>Science and Children</u>, Vol. 10, No. 1 (September, 1972), p. 26. By Permission.

game. What is the approximate population density per square centimeter of the stadium area in the drawing?

- Consider what the people may have brought to the stadium (cars, binoculars, rings, etc.). What approximate total wealth might the stadium population represent?

Some Facts are:

- Two-thirds of the stadium population is male.
- One-half of the male population is over 30 years old.
- One-half of the female population of the stadium is married.
- Forty percent of the stadium population walked to the game.

What time of day is it?

On this day, would you rather have the cola concession or the hot-dog concession? Why?

HOW DOES ONE BUILD A PYRAMID?*

So you want to build a pyramid. Simple, just pile up some rocks and there you are. This is not the way the ancient Egyptians did it. For thousands of years the pyramids of Egypt have excited the imagination of those who have been fortunate enough to view these enormous structures. Over seventy pyramids are known to exist in Egypt. How many more lie hidden under the sands of the desert, no one is sure.

How we think the Egyptians built Pyramids:

The largest pyramid, the Great Pyramid of Cheops was built by Cheops, the Pharoah of the 3rd Dynasty. This pyramid is 755 feet along one side of

* From Alfred De Vito and Gerald H. Krockover, Activities Handbook for Energy Education, 1981, p. 27-29, Reprinted by permission of Scott, Foresman and Company.

its square base and 481 feet high. This structure would cover 13 acres of ground or approximately an area covered by an eight city block square (or 64 square city blocks). The Great Pyramid contains about 2,300,000 blocks of limestone and granite weighing from 2 to 70 tons apiece and in size, averaging 50 x 50 x 28 inches. These blocks were stacked stairlike in approximately 200 steps to a height of a modern forty story building.

Little is known of the entire process of pyramid building. The process of cutting the stone blocks at the quarry, the transportation of these blocks, and the erecting of these monuments was such a common ordinary task for the Egyptians, they did not always consider it necessary to record how the job was accomplished. They just went ahead and did it. Most of the information we have is based on the study of the monuments themselves, especially those left unfinished when their builders died.

It is estimated that it took about 20 years to build a pyramid the size of the Great Pyramid. Using 2,300,000 blocks, how many blocks a year must be placed in position to complete the pyramid in 20 years? How many blocks must be positioned a day? In a twelve-hour day, how many blocks per hour would this be?

From an ancient tomb painting showing 172 men moving a 60 ton stone statue, it is calculated that eight men could move an average 2½ ton pyramid block. Thus, it is thought that teams of eight men were organized to work on individual blocks of stone. Various teams of eight men cut the block at a stone quarry. Another eight men transported the block to the pyramid building area. And still another eight men were responsible for raising the block to the required position on the pyramid. Various teams of eight men could complete this process for ten blocks in five months. If eight men could add ten blocks in five months to the pyramid, how many men would it take to furnish the necessary stone blocks for one year's work on the pyramid?

Why did the Egyptians build pyramids?

Many reasons are proposed for why the Egyptians built pyramids. Some more popular suggested reasons are:

The pyramids were built as tombs constructed in memory of individual Pharoahs. However, some people believe the pyramids were not tombs but temples for initiation into secret organizations.

The pyramids were built to serve as a means for recording the measures of the earth. The Great Pyramid is a carefully located fixed landmark. It served as a celestial observatory from which maps and tables of the stellar hemisphere could be accurately drawn. The Great Pyramid includes in its sides and angles the means for creating a highly sophisticated map projection of the northern hemisphere. It is, in fact, a scale model of the hemisphere, correctly incorporating degrees of latitude and longitude.

Some mathematicians have concluded that the Great Pyramid is a mathematical wonder. Whoever was responsible for building the Great Pyramid knew much about the dimensions of this planet. They knew much that was not rediscovered until the seventeenth century. The builders of the Pyramid could measure the day, the year, and the Great Year of the Precession. They knew how to compute latitude and longitude very accurately by means of obelisks and the transit of stars. The height of the Pyramid appeared to be in relation to the perimeter of its base as the radius of a circle is to its circumference. It is thought that the perimeter of the Pyramid might have been intended to represent the circumference of the earth at the equator while the height represented the distance from the earth's center to the pole (radius of the earth).

Circumference of a circle (C) = π x the Diameter (D) of the earth or $\pi = \dfrac{C}{D}$.

Divide the perimeter of the Pyramid (755 x 4) by twice its height (481 feet). The height equals the radius; two times the height equals the diameter. What value do you get? How does this value relate to the value of π ?

The truncated or flattened top of the Pyramid served not only as an observatory but also it served as a landing pad for space ships from outer space. Still other individuals believe the Pyramid was simply a stepping stone for Gods from heaven.

Some historians feel the Pyramids were built simply because man had reached the stage of development at which he was able to build them...much as modern man builds skyscrapers, the Eiffel Tower, etc.

Some modern day Egyptians believe the ancient Pharoahs in their wisdom built the pyramids anticipating the tourists that would come to Egypt to view them.

Egypt and the Pyramids

Egypt is a most peculiar country. It seldom rains there. The entire life of humans, animals, and plants depends on the water of the Nile River. This great river is almost wholly responsible for Egypt's economy. The Nile River through its more than 4,000 mile course to the Mediterranean Sea creates in the middle of a barren, sterile land a long, narrow, fertile oasis.

Each year the Nile floods its banks approximately from June through September. When the water retreats it leaves behind a layer of rich, black, silt. The Egyptians never had to worry about rain or fear exhausting the soil. The Nile provides irrigation and refreshes the soil.

The Egyptians had learned to determine the seasons of the year by the behavior of the river. The river has three seasons: "Flooding" from June through September; "Retreating of the flood waters," which begin in October and leaves the soil moist until about February; and "Drought," which lasts until June when "Flooding" takes place and the cycle once again repeats itself. From this cyclic observation the Egyptians created one of the first, practical calendars.

From these seasonal divisions also came social organization, for the river and its behavior determined labor assignments. During the "Retreating of the flood waters" men caught and stored the fast-retreating waters and planted in the mud. During the Drought they harvested and threshed. During the flooding stage they were available to cut and haul stones for the Pharoah's building projects. The pyramids were not built by slaves but rather by Egyptians who were forced to work for certain periods of time.

The selection of a place to build a pyramid was based on the following:

The pyramid, based on the Egyptian belief that the kingdom of the dead lay in the west where the sun sets, had to be located on the western bank of the Nile River.

It had to be near the cultivated land.

The ground upon which the pyramid was to be built had to be above the flood plain of the Nile River. If the ground rose up to form a plateau, a valley into the plateau could serve as a natural ramp for moving materials up the plateau.

The ground had to be almost flat. And, the ground had to be composed of a solid mass of rock to support the enormous weight of the finished pyramid.

It had to be reasonably near a source of rock from which the pyramid could be built.

How were the pyramids built?

Many explanations as to how the Egyptians moved the large, stone blocks have been proposed. Some say ancient priests of Egypt had the mental power to make the huge stones light and actually float them through space into position. Other people say the Egyptians had anti-gravity machines which made lifting huge stone blocks a simple task. Most people, however, think the job was accomplished with nothing but primitive tools, patience, and unlimited manpower.

Pulleys were unknown in ancient Egypt. In quarrying and building, workmen used copper chisels and possibly iron tools, as well as flint, quartz, and diorite rock pounders. The only additional aids used were large wooden crowbars and for transportation, wooden sleds and sometimes rollers.

To quarry the rock from the hillside, the Egyptians chipped away vertically with a wooden mallet and a copper chisel, which must have been highly tempered by some method unknown today. The Egyptians cut a groove in the rock where they wanted it split. They then drilled holes along the groove, beat wooden plugs very tightly into the holes and poured water over them. When moistened these wooden wedges expanded to crack the rock. Sometimes fires were built along the grooved lines, and water poured on the heated stone to obtain a clean break in the rock.

Even with adequate though simple tools, patience, and unlimited manpower, the great organizing ability to start and finish such a project as a pyramid had to require a very special skill. It would take a genius' talent to plan all the work, to lay it out, to provide for emergencies and accidents, to see that the men in the quarries, on the boats, and sleds, and in the masons' and smithies' shops were all continuously and usefully employed, that the means of transportation was ample, that the food supply did not fail, that the water supply was adequate, and conveniently available, and that the sick relief were on hand and available for replacement. It is estimated that as many as 40,000 skilled workers lived permanently on the spot. Feeding, housing, and clothing these workers was a tremendous undertaking.

As to how the Great Pyramid was actually constructed, there are many different opinions. For the lack of such equipment as steel cranes and derricks to lift and swing heavy blocks of stone, the Egyptians would have had to construct a ramp or ramps to raise heavy stones to the required level. Building the ramps was almost as great a task as building the pyramid itself. Several ramp constructions are suggested by Egyptian historians. Some of these are:

> As the pyramid rose, the builders raised an earthen mound on all sides of it, with one or more long ramps for hauling up the stones. As each row of blocks was laid, the mound and the ramp was raised another level. When the pyramid was completed, all this vast amount of earth had to be hauled away.

> By using the building itself as a ramp, the Egyptians dragged the stones up the pyramid's own spiraling outer wall. This would enable the builders to fill in the core as they went up and finish the casing as they came down.

> As the layers of the pyramid began to rise, four ramps made of stone rubble and mud were also built. Each ramp started at one of the four corners and rose to the unfinished level. The slope of the ramp was always maintained at 15 degrees.

Which ramping technique would you use? Having only simple machines such as the roller, the lever, and the inclined planes (the Egyptian made little use of the wheel), what techniques would you use to raise these large stone blocks into position? What happens to the length of the ramp as the elevation of the pyramid increases and the slope of the ramp is maintained at 15 degrees?

The limestone blocks and rubble for the interior of the pyramid appear to have been cut from local outcrops. This was dragged directly to the building area on sleds. To make the sledding easier, a liquid (probably milk, the fat content of which makes it a better lubricant than water) was poured on the ground in front of the path of the sled. Finer limestone had to be rafted from across the Nile River almost 20 miles away. Granite used for the linings of chambers was floated on barges down the Nile from Aswan almost 500 miles south of the Great Pyramid.

Assume that you are an ancient Pharoah and that you want to build a pyramid. Where would you build one, and how would you build one?

- Using your state map pick the best spot within your state for the location of your pyramid. You are restricted to building your pyramid using only tools that ancient Egyptians used, for example, rollers, the lever, and the inclined plane. Review the considerations the Egyptians concerned themselves with in the selection of their pyramid sites.

- Design your pyramid designating the size of the base, the height, and the slope of the sides. Make a scale model of your pyramid stating your technique for proceeding with the building of your pyramid. What would you do first, second, third, etc.? Your model can be constructed from wood, cardboard, sugar cubes, etc. Design the internal secret passageways and hidden rooms. How would you design the pyramid so that thieves could not break into the pyramid?

THE SILENT GUARDIANS OF EASTER ISLAND*

History may be recorded in many ways: it could be a Revolutionary War rifle ball embedded in a tree at Gettysburg, Pennsylvania, a Roman road sprawling across the countryside of Europe, or the Great Sphinx overseeing portions of the Nile River.

History is punctuated with structures that baffle investigators such as anthropologists, geologists, historians, physicists, and curious observers who wonder how and why they were constructed. The magnitude of many of these structures boggles the mind. Ancient people, located in various places of the world, constructed a variety of massive structures. A few of these are:

. A peculiar drawing was carved high on the red wall cliffs south of Lima, Peru. Viewed from the sea, this 820 foot high carved drawing can be seen twelve miles away. The carved figure resembles a gigantic 3-pronged, fish spear. Oddly enough a long rope was found hanging down from the central prong of the carved spear. What purpose did the carved drawing serve? What was the rope used for?

. Another interesting structure whose construction has puzzled archeologists is a series of circular setting of large standing stones called Stonehenge. Stonehenge is located on the Salisbury Plain in England. An innermost ring consists of five pairs of gigantic stones each bridged with a horizontal piece. The vertical stones weighed some forty tons apiece.

. Still another impressive engineering feat are the Egyptian tombs built in pyramid form. Stone blocks averaging two and one half tons each were used to build these pyramids. Some pyramids were made up of two million

---

* From Alfred De Vito and Gerald H. Krockover, Activities Handbook for Energy Education, 1981, p. 25-27, Reprinted by permission of Scott, Foresman and Company.

blocks. Occasionally, larger stone blocks were used - some as heavy as fifteen tons apiece.

. Not far from the sea, in the Peruvian spurs of the Andes Mountains lies the ancient city of Nazca. A nearby valley contains a strip of level ground some thirty-seven miles long and one mile wide. If you fly over this territory you can make out gigantic lines laid out geometrically. Some of these lines run parallel to each other, others intersect or are surrounded by larger trapezoidal areas. Some archeologists say that these lines represent ancient Inca roads. If these are roads, they lead nowhere. They just end sharply. Some say these lines represent some expression of religion or perhaps even a calendar. Other individuals have interpreted this construction as a landing field for ancient aircrafts that may have visited this planet from outer space. No one knows.

Perhaps the most interesting and challenging mystery of the past lies hidden on Easter Island.

Easter Island is the loneliest inhabited place in the world. It is located in the South Pacific Ocean about 2,000 miles west of the Chilean coast. It covers only fifty square miles and its population numbers about three hundred people.

Ancient people who once lived on Easter Island were called "long ears". They pierced their ears and dangled heavy weighted objects from them to stretch the lobes. Eventually their ear lobes reached down to their shoulders. Descendents of these ancient people still live on Easter Island. However, long ear-lobed people are not visible anywhere - only short ear-lobed people can be seen on the island.

Stranger than the disappearing long ear-lobed people is the silent army of stone statues that stand watch over the island peering out to sea. There are over six hundred statues on the island. These giant statues are images of the long ears themselves. Every statue is exactly the same. All statues have the same expression on the face. They are made of the same extremely, tough, grayish-yellow black-grained stone. When this stone is struck with a steel axe, sparks fly. Every statue has a flat cut-off foundation. This cut-off is where the stomach ends and the legs begin. With one exception, all the statues are male figures. These statues are like nothing else found anywhere in the world.

    Each statue weighs an average of fifty tons (equal to about 25-30 medium sized automobiles) and ranges in height from 30 to 69 feet tall.

    The rock supply from which the giant statues were cut is the steep side of a water-filled, dead volcano. The volcano has been cut up like it was made of soft dough. Hundreds of thousands of tons of stone has been cut out and tens of thousands of tons of stone was carried away. In this gaping hole lie more than one hundred fifty gigantic stone giants in all stages of completion -- some just begun and some just finished. The carved figures were cut in vertical and horizontal positions. It did not seem to matter to these ancient rock carvers in what position they worked. At the foot of the volcanic mountain stand finished stone men side by side like they had just come off an assembly line.

The major portion of the stone cutting and polishing of the statues was done at the volcano. The only stone carving that was completed after having moved the statue to a desired spot on the island was the carving in of the eyes. It was as though the stone carvers did not want the stone giants to see where they had been or where they were going. After being carved and polished these statues (fifty tons in weight and forty feet tall) were moved all over the island. Some were moved as far as ten miles away over rough and hilly ground. Each statue was not only moved but when it reached its destination, it was lifted up and placed on the top of a platform that was built up some six feet above ground level. The tremendous effort did not end there. A gigantic head decoration weighing from two to ten tons (This is almost as much as the weight of two elephants. The average volume of the head decoration is approximately two hundred cubic feet.) was placed on top of the silent stone giant's head. This decoration was similar to the hairdo worn by male natives of Easter Island. The color of the head decoration must have been important. The giant head decorations were cut from still another volcanic quarry which contained only red rock. This red rock quarry is located seven miles from the supply of rock used for the body of the statues.

Some facts:

- It would appear that the ancient natives had no knowledge of the wheel, no cranes, no motorized vehicles of any kind, no metal tools, and no roads as we know them.
- Present observations of this area would lead one to infer that it was as though all the ongoing work had stopped abruptly. Thousands of primitive unpolished stone picks still lay in the open. All stages of carving were visible. Some completed, some half completed, and some just started. It did not look as if they were planning to stop the business of producing stone giants.
- Some head decorations were found in the shallow edges of the island's water.
- Bones of dead natives were found at the feet of these giant stone statues.

How was this all accomplished before the age of modern machinery? Is this possible without the use of machinery? These statues were not made by wood carvers who simply change from carving wood to carving rock. These statues were cut from rock by people with experience in cutting and moving these giant statues.

How did these ancient natives:

- carve these huge statues out of this tough rock using only crude stone picks?
- move these huge blocks of rock down the mountain side and then over land, sometimes miles?
- get the enormous hair decorations on top of the vertical statue?

Why did:

- they polish these stone statues before moving them?
- the ancient natives want long ear lobes?

The present day natives on Easter Island when asked how they think the statues were transported simply state, "They went of themselves." What do you think?

VITA VISTA REVISITED

Approximately six square miles (about 3900 acres) of unused rangeland ("A"), about two miles east of town, is now available for the town's use through a donation from one of its citizens.

Study the background information for the town of Vita Vista, then list some possible uses for the rangeland by the town. From this list select your best choice.

Background Information:

- The population of Vita Vista is 150,000 and rapidly increasing.
- Thirty-five miles northeast of town out on highway 94 is located the town of Grande Vista with a population of over 500,000.
- There is no town southwest of town for a distance of 150 miles.
- Increased population growth is resulting in greater demands for housing, jobs, additional city services such as water, sewage, garbage disposal, schools, etc., plus recreational services.
- The town is growing and will soon extend beyond its present boundries.
- The town's people would like to keep the town from growing too rapidly and they would like the town to maintain its natural beauty.
- The areas north and west of town are highly productive farming areas.
- Ranchland area labeled "A" could be farmed but at the present time it is being used as a range for cattle.
- The mountain area west of the reservoir is heavily forested with an excellent stand of hardwood trees.

- The Blue River is unpolluted and serves as the town's drinking water supply. It is also used for irrigation.
- The Blue River flows swiftly and while large boats cannot travel on it, logs could be floated down it.
- The present sewage treatment plant and garbage disposal facilities are operating at maximum capacity.
- The mineral zincite has been mined in the hills east of town. The extent of these deposits is unknown.

Given the facts about Vita Vista, what are your suggestions as to the possible considerations for the use of the donated land?

What questions, which if you had the answers to them, would assist you in making improved suggestions relative to use of the vacant land?

Given the facts, what is your best choice for usage of the donated land?

309

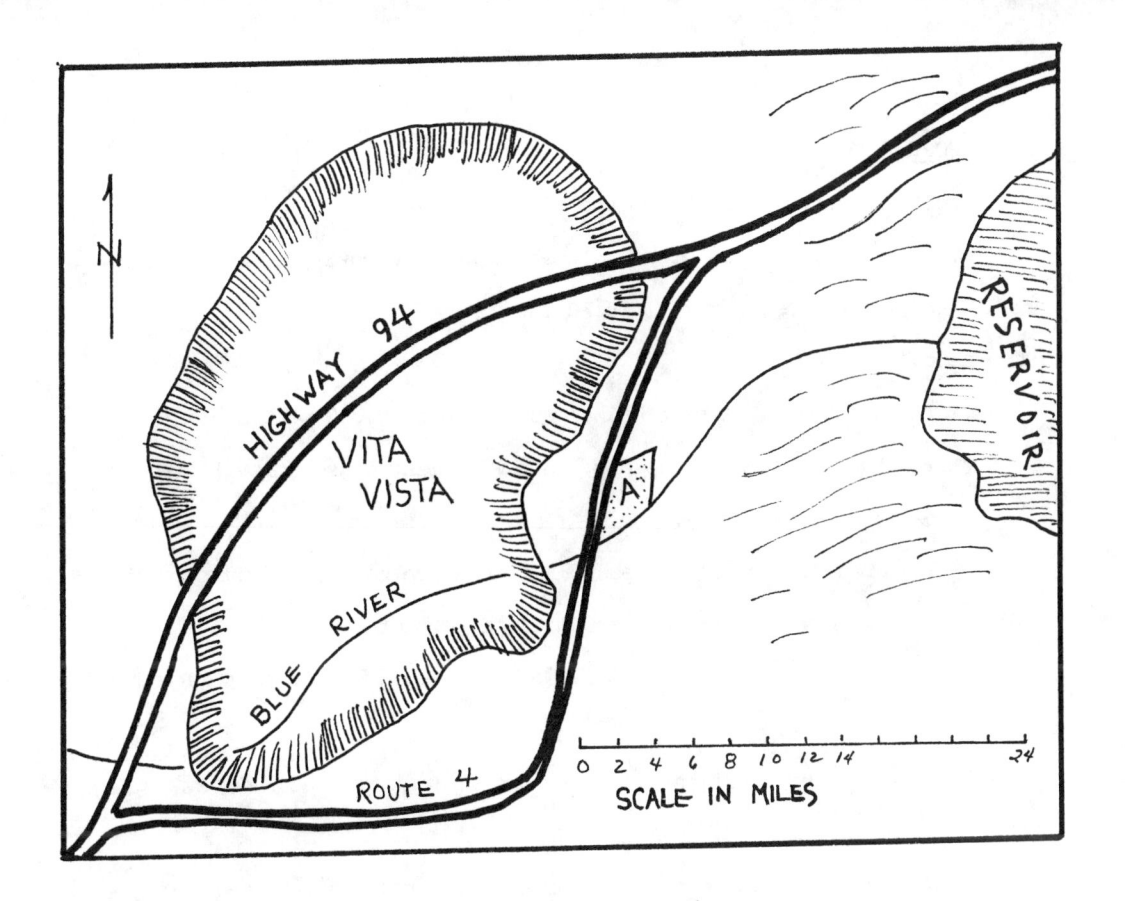

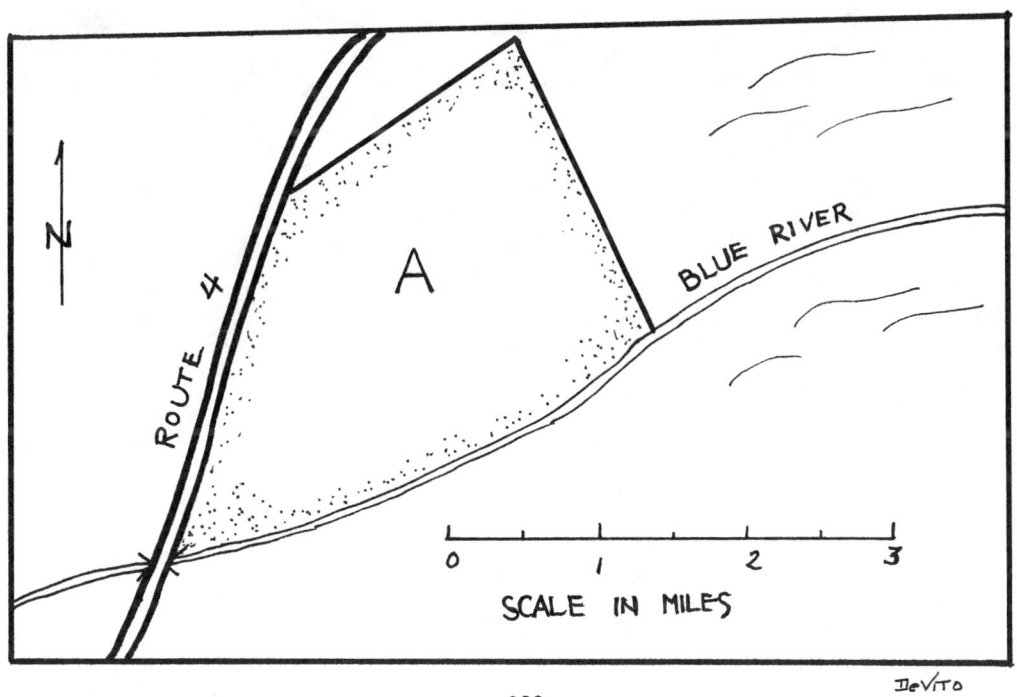

CENTERPORT, USA

Problem: Where do we build the elementary school? The town of Centerport's population has grown to where a new elementary school is needed. Where is the best place to build it?

Some facts are:

- The town of Centerport has a population of 9,000 people.
- The town is growing rapidly. A new elementary school will need to be built immediately.
- The area of Centerport marked A thru Z are new homes under construction. They are all sold and will be completed soon.
- The town's boundries are well defined and marked by four surrounding communities.

What factors are the location of the future elementary school dependent upon?

Each child is assigned a role, for example, real estate developer, mayor, president of the local bird watcher's society, librarian, farmer, garbage man, etc., etc. Assuming one of the roles, each individual will work out a solution. After each individual has worked out a solution that reflects his role's interests, a town meeting is called to determine whether a single solution can be agreed upon.

Note:

- Land can be purchased or sold.
- Portions of town can be relocated.
- The teacher can assume any role in the discussion.

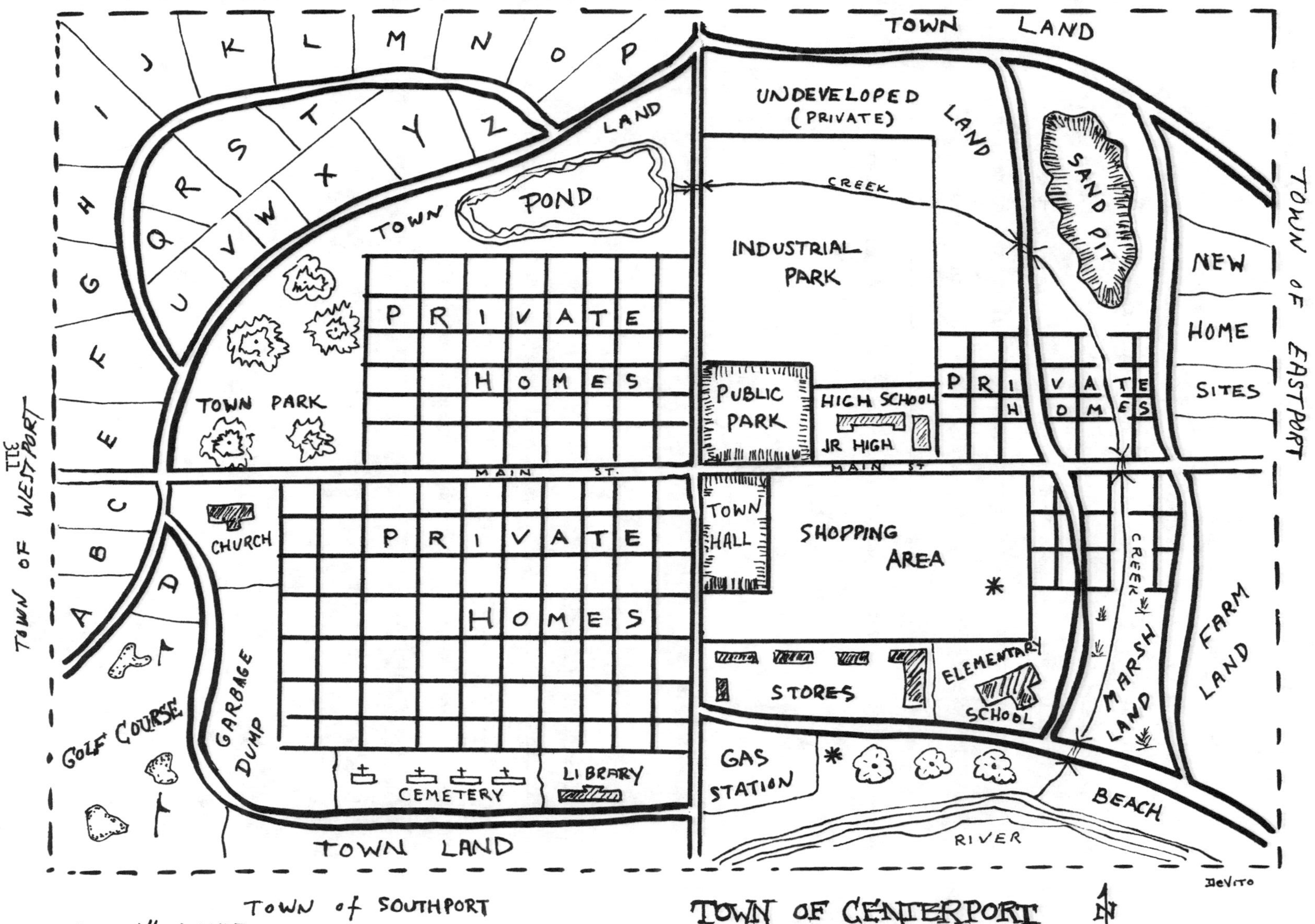

TOWN OF CENTERPORT

SCALE 1" = 1 MILE

ANOTHER FAST FOOD CHAIN

The town of Centerport has two, fast-food restaurants located at each asterick (*) on the map. These are the only restaurants in town. A new, fast-food chain is planning to locate somewhere in the town. What factors or conditions are relevant to selecting the best location? Survey the map of Centerport and on the unfinished map make your selection for the best locations by noting on the map your first, second, and third choices for the fast-food restaurant's location. Assemble as much data as you can to defend your selected sites. Compare your choices to your classmates.

THE BEST GARBAGE TRUCK ROUTE

Determine the most preferred garbage truck route for collecting Centerport's garbage. Garbage is collected twice a week from everyone in town. The town has two garbage trucks. What concerns must you take into account relative to establishing the most preferred routes? Trace your recommended route. How does your choice of a preferred route compare to other choices made by your classmates?

TOWN OF CENTERPORT

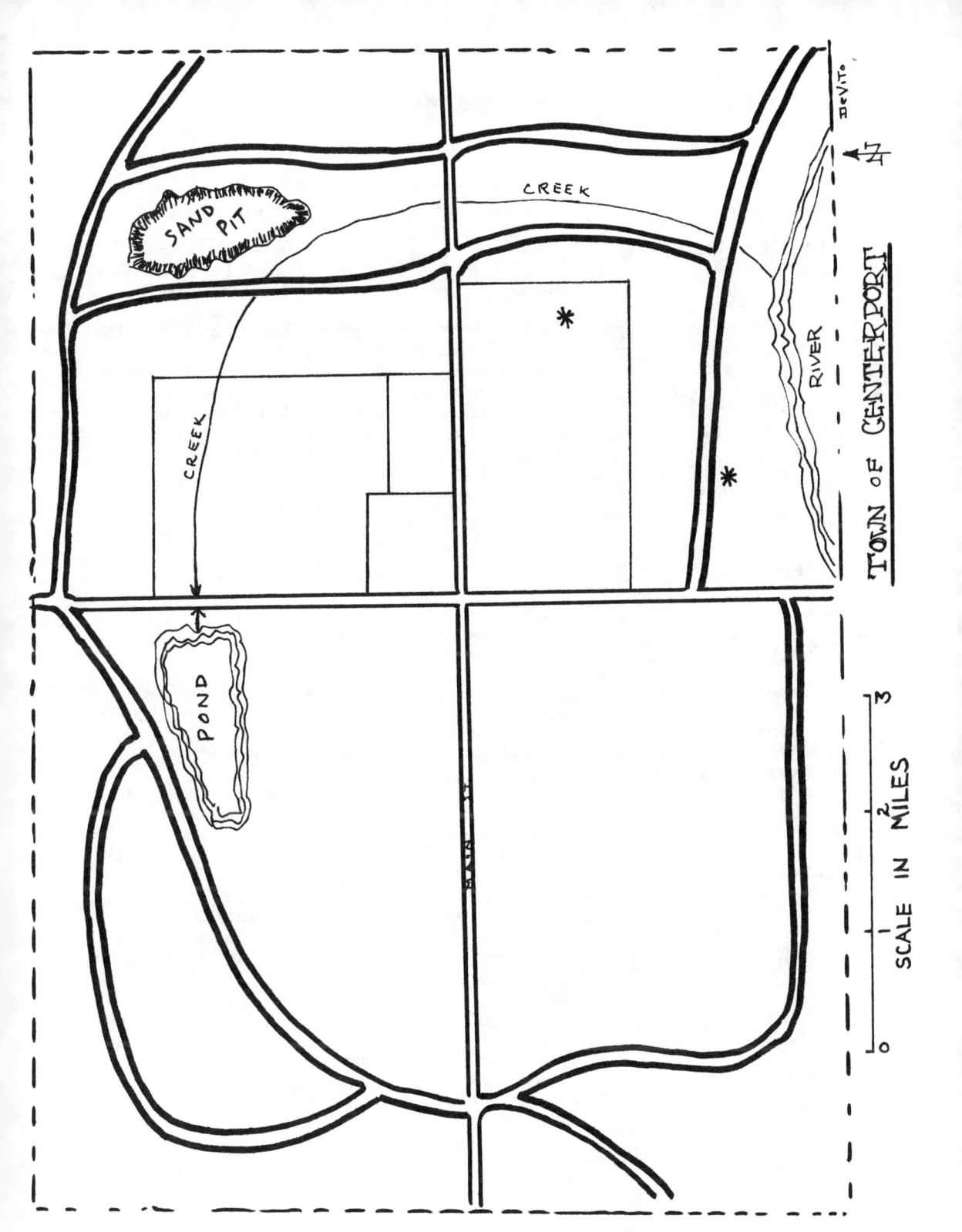

Casino

CHARACTERS

Leonardo da Vinci
Lewis and Clark
Julius Caesar
The Bionic Man
Laurel and Hardy
Albert Einstein
Snow White & the
    7 Dwarfs
General Custer
Amelia Earhart
Frankenstein
The Three Stooges
Noah (and the ark)
Adam and Eve
Huck Finn
The Flintstones
Me, Myself and I

SITUATION

In a Laboratory
State Fair
Halloween Evening
World Series Playoff
Invasion by Martians
First day of College
An Elementary School
A Jury Trial
A Circus
Lost in the Catacombs
A Tour through the
    Statue of Liberty
Stranded on a Deserted Island

* The die can be constructed from a Manila folder (see cubic layout) or a cube of wood.

1. Select from each category any six items. Number each group of six, one through six.*
2. Roll the die five times, once for each category.
3. Each roll of the die will provide you with a number from one to six.
4. The number which appears on the die should correspond to a number in each of your categories.
5. After you roll the die five times, assemble the five items as one composite list.
6. Using these five items, weave them into a creative story.

MESSAGES

Cheating doesn't pay.
The good guys always win.
Jokes may get out of control.
You learn from your mistakes.
Better safe than sorry.
Haste makes waste.
Waste not, want not.
Save for a rainy day.
A penny saved is a penny
    earned.
A fool and his money is soon parted.
Many are called and few are chosen.
An apple a day keeps the doctor away.
Nothing ventured; nothing gained.
Honesty always (sometimes, never)
    pays.
A stitch in time saves nine.
Laughter is good for the soul.

PLACE

Mars
North Pole
The Sahara Desert
New York City
Inside of a Pumpkin
Moon
Center of the Earth
The White House
In a Space Capsule
Under the Ocean
The Jungles
Death Valley
Disneyland
Cape Kennedy
Inside a Whale
The North Pole

TIME

The Dinosaur Era
Western Frontier Time
Ancient Grecian Time
Civil War
2000 B. C.
The San Francisco
    Earthquake
The Gold Rush Days
Westward Expansion
The Era of Pyramid
    Building
Ancient Rome
The Glacial Period
    in the U. S.

WHERE DEPENDS ON WHAT?

You are attending a school banquet. There will be a very important guest speaker. Dinner will be buffet style. The banquet will last about two hours. There will be at least 250 people in attendance. Where in the banquet hall is the best location to sit?

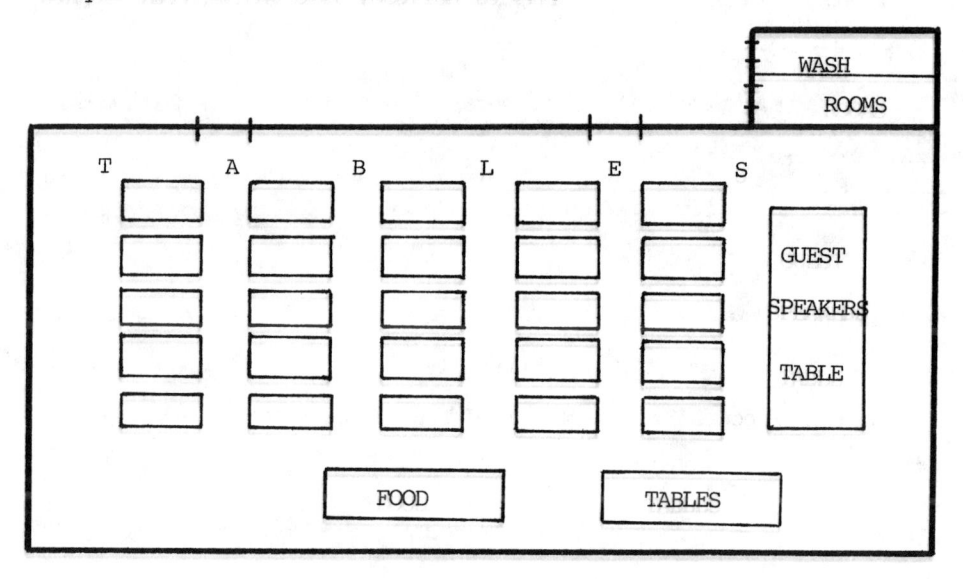

Seating preferences could be based on:
- proximity to food
- ability to hear the speaker
- ability to exit in case of a fire
- ability to see the speaker
- air conditioning (or if winter - heat)
- incoming sunlight
- ability to exit at the end of the session
- how far from the rest rooms

Where is the best overall place to sit considering all or as many of the above conditions?

AND NOW WHAT? NO. 1

Originally there were two million acres of Wisconsin prairie land. Now there are less than 20,000 acres.

Once California had two million acres of redwoods. Now there are less than 50,000 acres.

Once there were five billion passenger pigeons in these United States. Now there are none.

Once there were fifty million bison. Now there are about ten to fifteen thousand.

Cars, trucks and buses in the United States now add 85 to 100 million tons of pollutants to the atmosphere annually.

Factories now pour 22 million tons of contaminants into the air each year.

Electric power plants now add 15 million tons of pollutants to the air each year.

Heating of homes and buildings add another 8 million tons of pollutants to the air annually.

And now what?

- Can this continue?

- What alternatives are available to us?

- What do you suggest?

AND NOW WHAT? NO. 2

Arizona authorities, fearing a new threat to their fruit crops, have launched a campaign to rid the state of the Chi-Chi-Chuca birds that descended from a pair of zoo escapees. These rare birds are not native to the area, but they seem to be adjusting nicely to the Arizona setting.

About 100 of the tiny blue-green, Southeast Asian birds - for which even zoos need special permits - have been trapped so far in nets.

The Chi-Chi-Chuca bird, so named because it screams "chee, chee, chee" as it chews the soft fruit it punctures with its sharp beak. A spokesman reported that the state department of Food and Agriculture wants them captured before they migrate into other orchards plus migrating into the tomato fields to the east and west.

The first Chi-Chi-Chuca bird apparently flew out of an aviary in the Tucson Zoo in 1978. Since then they have bred and some have been found several miles from the zoo.

And, now what?

- Should we stop raising fruit and starve them out?
- Shall we spray them to death and perhaps ourselves included?
- What do you suggest? What is your plan for controlling the Chi-Chi-Chuca birds?

THE RESPONSE SHOULD BE...

Peking - The year of the rat has arrived ahead of schedule in China.

Large parts of the nation of China are plagued by billions of rats that bite farm animals and chew up more than 20 million acres of prime grain fields, forest and pasture. The epidemic has reached almost astronomical proportions in central China, where one province claims to have a rat for every 70 square feet of farmland. Two state farms in the south saved their crops by killing 106 tons of rats last year. But peasants in the northeast grasslands were less fortunate and expect their hay harvest losses to reach about $400 million.

Some facts:

- In the first half of this century bubonic plague killed a million people.
- Chinese zoologists say this is the peak year for rat production because of the drastic decline in their natural enemies - cats, weasels, owls, and snakes.
- The rats appear to be winning the survival race. Their predators have died from feeding on rats that had eaten poison. So the rats' enemies are quickly vanishing.
- The cycle of nature favors rats because they propogate at much faster rates than their predators. A mother rat is capable of producing up to 100 babies annually.
- Rats are benefiting from the ecological havoc caused by Chinese peasants who poach weasels for their pelts and illegally cut down forests for building materials, thus destroying the habitat of owls and snakes.
- Many cats were destroyed in the 1970 because they were considered property of the rich.
- In south China rat meat is regarded as a delicacy.

What should be done?

Can you suggest a plan which may solve the problem?

What do you recommend? Defend your position.

IN YOUR OPINION...

Crime tarnishes image of Pleasant View Park (News Heading)

Violence has chilled Pleasant View Park, just as its daffodils and tulips hit full spring bloom.

Sarah Stringlewind, 17, of Butternut County was shot to death and her male friend wounded on a sunny Sunday, March 6 in a busy section of the park.

Three days later a park attendant was mugged - the fourth mugging in the park this year.

In addition, police now believe a suspect in 17 assaults in and around the park since February 7 has "settled" in the nearby neighborhood.

The park is 115 years old. Crime now is no more prevalent than usual. Pleasant View Park is one of the nation's largest urban parks with 1,017 acres of land, 3½ miles of gently curving roadways and some 150,000 trees. The park drew up to 20 million visitors last year.

There were nine murders in New York's 840 acre Central Park last year which drew 14.2 million visitors that year. That year Pleasant View Park had no murders.

- Should we close the park?
- If we decide to leave the park open, what changes might you suggest?
- In your opinion. . .

PRESCRIPTION FOR A REEF

In the early 1960s observers noted that the starfish called crown-of-thorns starfish (Acanthaster planci) was undergoing a population explosion along portions of Australia's Barrier Reef. Soon after, the reef began to disappear. The reef was thought to be almost indestructible. However, it was found to be appetizing and was being devoured by the crown-of-thorns starfish which quickly gained the title of "Reef Killer."

But, the crown-of-thorns had his problems too. It was thought that the crown-of-thorns had just one effective predator, the marine triton. But it has been discovered that the painted shrimp plus two other marine invertebrates, the xanthid coral crab and the snapping shrimp, were attacking crown-of-thorns. It seems these crustaceans apparently can identify the presence of the starfish just from a solution of the sea water in which the starfish lives. Apparently, the crown-of-thorns starfish when eating, turns its stomach inside out, spreading digestive membranes that break down the coral's soft tissue.

Several years ago geologic coring of reefs uncovered repeated layers of starfish skeletons, evidence that these population explosions have taken place every 200 years for several millenia.

- Do we have a problem? If so what?
- Scientists think the crustaceans may have developed their response to the mysterious starfish substance as one way of helping to control the crown-of-thorns. What do you think?
- Should we attempt to preserve the reefs? Should we try to regulate the crown-of-thorns? Or, should we just let things take their course?

STONEHENGE - WHY AND HOW?

In all parts of western Europe, Britain included, the landscape is marked with tombs built with large upright slabs of stone roofed over with an even larger slab and covered with earth. Some are single standing stones, some are lined up in rows, and some are in circle forms called henges. Wind erosion in some cases removed the earth covering and exposed the stone tombs.

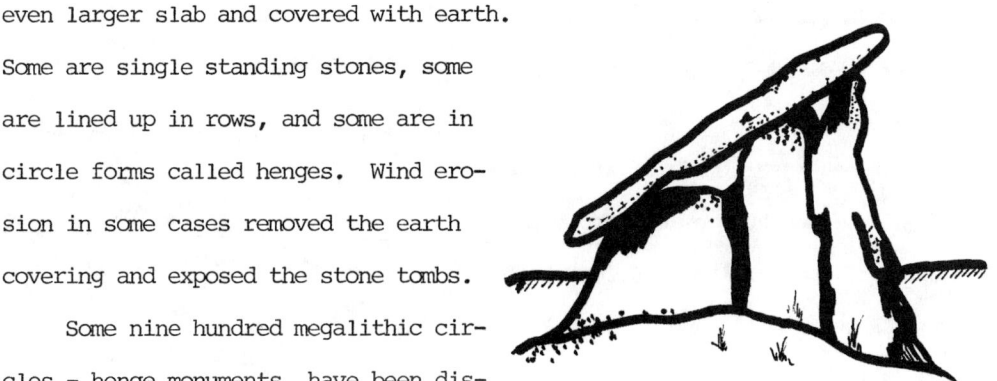

Some nine hundred megalithic circles - henge monuments, have been discovered in the British Isles. Not all are alike, the same age, or serve the same purpose. Stonehenges are generally megalithic, meaning constructed of large stones. Stonehenge has been standing on the Salisbury Plain in southern England for thirty-five centuries. During this time men of each age have seen in Stonehenge whatever purpose they had reasoned it to be. Some historians in the 12th century claimed it to be a group of memorial stones raised by Merlin the Magician for a British cemetery for the fallen dead of various wars. Later in the 17th century Stonehenge was thought to be a Roman temple dedicated to imperial grandeur. During the reign of Charles II, this monument was thought to be a place where early kings were crowned. During the period of the French Revolution this stone structure was reasoned to be a temple directed by British Druids against the Roman conquest. Recent interpretations consider Stonehenge as a scientific observatory for the prediction of solar and lunar events. Some have wished to elevate Stonehenge to the level of an astronomical observatory. However, evidence for this conclusion is doubtful. Nonetheless, Stonehenge does seem to be aligned with celestial indicators of the four main seasons: which is to be expected of farmers, herders, and sea navigators.

Stonehenge is made of stones some of which weigh 50 tons and whose dimensions are approximately 30 feet high, about 3 feet thick, and roughly 7 feet wide. Thirty smaller stones about 26 tons each surround these larger stones. These thirty, smaller stones were originally erected in a circle with a 48½ feet radius. These were erected in an upright position and capped with a lintel stone to form a ring of archways. All of this is surrounded by a circular ditch about 320 feet in diameter. This ditch is about 35 feet wide and varies from 2 to 60 feet deep.

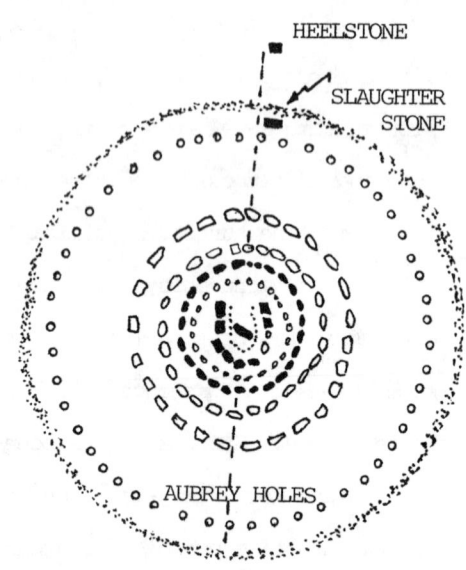

Stonehenge stones have been established by means of radio-carbon dating as being older than the pyramids of Egypt, and even older than the first cities of Mesopotamia. They have been identified as spotted igneous rock native to the Salisbury Plain Area. These stones are as hard as quartz and harder than steel. It is not easily pulverized under blows from a hammer stone, which can only remove 5 cubic inches per hour.

- Was Stonehenge built by magic?
- Is Stonehenge the remains of monuments built by ancient outer-space visitors?
- Or is Stonehenge the visible remaining artifact of a primitive people; their technology, their society and their expression of mortality?

What do you think?

If Stonehenge was built by magic, the huge stones may have been put into position by levitation.

If you think ancient outer-space visitors were responsible for its construction, it may have been accomplished in a variety of ways, for example, laser beams, etc.

If you think primitive, pastoral people built Stonehenge, how do you think they erected such a monument?

WHAT EVER HAPPENED TO THE DINOSAURS?

Dinosaurs, during the Mesozoic Era of geologic time, evolved into two main groups, distinguished by different pelvic-bone structure. One group had pelves in which the lower bones were long and parallel. Most dinosaurs in this group ate plants and traveled on all fours. The other group had its lower pelvic bones angled and joined in a strong arch that could support all the body's weight on two legs. Some of this group were plant eaters, but most were huge, swift, aggresive, meat eaters.

The Dinosaur brain occupied a tiny recess at the rear of the skull and its primary function was probably to work the jaws and receive general sensory impressions as to the location of food and possibilities of danger. One species of dinosaur had three brains - one located in the head and two located in the spinal cord to control leg and tail movements. Despite three brains, the combined effectiveness of these brains would have been inferior to those of a present-day domestic kitten.

Dinosaurs, who at one time were distributed worldwide, were extinct seventy million years before the first man appeared on the Earth. Their remains have been found on every continent. Many ideas have been put forth to explain their disappearance from the face of the earth.

Some think:

Climatic changes such as widespread glaciation caused widespread modifications in plant life, so that the herbivorous dinosaurs were deprived of their food. As the plant eaters decreased so did the meat eaters who fed on the plant eaters.

Some think:

There was a large-scale epidemic that was the result of the sea dropping its level and causing land bridges to be exposed. Thus, animals from the various continents intermingled. This intermingling caused the animals with no immune protection against newly introduced diseases and no defense against unknown predators, to simply die out.

Some think:

Sixty five million years ago, a giant asteroid roared through the atmosphere and hit the ocean water with a shock 1,000 times greater than the largest earthquake ever recorded. This would quickly darken the sky as debris from the exploded asteroid and dust churned up from the sea floor. The sky would turn black as this material spewed up into the stratosphere. This material would spread into a dense layer and obscure the sun.

Superheated, sea water thrown up into the sky would come down as warm rain. Lastly, these dinosaurs would be hit by a 500 feet wall of ocean water. This tidal wave could race around the earth within twenty-seven hours. Most of the lowland dinosaurs may well have drowned ending their 140 million year reign on this earth.

Some think:

Racial old age killed off the dinosaurs. A group of animals, after a period of successful adaptations, finding itself in a comfortable environment with plenty of food, and a lack of serious competition, does not need to change in order to survive. The hormones of this race, deprived of their true functions, begin to indulge in useless experimentations, for example, size in which case the reproduction cycle takes longer and fertility of the dinosaurs is diminished. The species loses its vitality and capacity to adapt. The race becomes decadent, if unable to adjust, the race becomes extinct.

- Does any one explanation sound possible? Or, might a combination of two or more explanations explain what happened to the dinosaurs?
- What do you think? What explanation do you have?

ANY UFOs OUT THERE?

Is space travel really here? Do we have visitors from outer space checking us out? If so, are they friend or foe?

It has been calculated that there could be 50,000,000 civilizations in our galaxy alone and perhaps 130 billion possible cultures in the entire cosmos.

It has been shown that through the use of staged fission or fusion propulsion systems would be capable of making trips between stars in less than an earth - traveler's life span.

There are 90,000 reported cases of UFO sightings. This number represents 25 percent of all those reported. The other 75 percent have been disregarded because something was unacceptable with the evidence. Some UFO observations are further confirmed by traces of their landings. These observations number approximately 15,000. This evidence might be indentations or impression rings left in the ground from the mass and shape of the UFO and accompanying radiation which has permeated the soil. Sometimes the UFOs caused such high winds on take-offs and landings that tree branches were broken.

UFO sightings are not new events. One UFO was observed flying over the United States in the year 1896. One was sighted in Portugal in 1917. However, the heavy influx of UFO observations has occured during the past 35-40 years. Upper northern California reports the highest number of sightings in the world.

Along with visual observations more tangible pieces of evidence have been accumulated. In 1962 a piece of a UFO was shot off over Washington. It was found to be composed of magnesium orthosilicate, with some small, round inclusions. Another UFO piece recovered in Sweden was a chunk of tungsten carbide, one of the hardest materials known. Some fragments recovered from Brazil were made of ultra-pure magnesium.

A composite model of viewed UFO aliens as reported by the majority of people who have seen the visitors from outer space describes them as hairless, chalky-skinned beings, four to five feet tall; with the teardrop portion of the eye extending all the way to the back of the head; and with no external ears visible.

Some additional facts:

- UFOs always seem to disappear. None stay around in any one place for very long.
- Photographs have been taken. UFOs generally appear disc-shaped, thirty feet in diameter with a small tower on the top portion.
- UFOs have been observed to accelerate rapidly, to almost float in the atmosphere, and to defy the laws of gravity.
- At night they appear to possess a luminous glow.
- Polarized light, in the vicinity of UFOs, is influenced as if by a magnetic field.
- Some UFO viewers actually claim they have been taken aboard UFOs. So far as we know none have been kidnapped; all have been released without any permanent harm and no ransom has ever been demanded.

What do you think:

Are UFOs just the result of overactive imaginations?

Are there such things as UFOs?

Are UFOs from this planet or some other planet?

What explanation can you provide that would either support or refute the existence of UFOs?

CAUSE AND EFFECT OR "WHEN IS IT MY TURN?"

The United States Government plans to take 82 million acres of farmland out of production this year.

Farmers have agreed not to plant 53 million acres of farmland in return for government surplus grain which will be credited to them under a program entitled Payment-In-Kind or PIK. It is anticipated that some 29.3 million more acres will be idled through other programs.

This action in response to government's programs could wipe out federal crop surpluses this year. This will result in higher farm and consumer prices.

It seems like all the problems are solved, but...

- Farm supply companies will lose $10.5 billion.
- Fertilizer companies will lose $2.7 billion.
- The farm equipment and machinery companies expect to lose $2.5 billion.
- Fuel and lubricant companies speculate they will lose $13 billion.
- The seed companies believe their losses will be approximately $770 million, and
- The chemical industry feels their losses will be in the range of $872 to $875 million.

Some think:

- This action will wipe out surpluses earlier and farmers will begin producing sooner and demand for machinery and other supplies will rise.
- In the long run it will be good for everyone.
- This action will cost a 7 to 10 percent drop in a $196.5 billion dollar industry. This could have a serious impact.

What do you think?

ANCIENT ORIENTAL MARINERS -- OR "WHO DISCOVERED AMERICA FIRST?"

Who did discover America? Some say Viking mariners arrived four or five hundred years before Columbus; and several hundred years before them perhaps Irish monks visited America. And some others even proposed that ancient Phoenicians at the time of Christ arrived before all of them. It is believed that our eastern shores were invaded by many in a variety of ways and times.

Some historians believe our western coast was also invaded, perhaps prior to all the eastern invasions, by Asians - Chinese, maybe Japanese or even Afghanians. The Chinese and Japanese have a long history of ocean travel. The Chinese developed the balanced rudder nearly a thousand years before Europeans did; they sailed ships constructed with watertight compartments; and used compasses centuries before such devices were known in western ships.

Ships were observed in western waters as early as 1540 when Francisco Coronado's patrols discovered two small vessels anchored off the coast of the head of the Gulf of California. These sailors indicated by signs that their homeland lay to the west in the direction of China or Japan. Thirty-three years later, a flotilla of eight ships were sighted sailing in the ocean off the west coast of Mexico. This was four years before the first non-Spanish European vessel first entered the Pacific Ocean.

Did ancient mariners visit our west coast?

Some facts:

- Buried implements, thought to be of Asian origin were uncovered in the Pacific Northwest.
- A large boulder at the joining of the Ada and Snake River in Idaho is believed to be inscribed with primitive Japanese letters dating

from around the ninth century A.D.

- Primitive Chinese picture writing have been found in Nevada's Valley of Fire.

- On the west coast of Washington State's Olympic Peninsula a large number of high-carbon steel blades - knives and chisels have been found in an old Indian village. Chemical analysis of these blades have shown their steel to be a kind of steel manufactured in Japan as early as the eighth century A.D.

- In a fishing village in Ecuador its inhabitants make a pottery that closely resembles pottery manufactured on the Japanese island of Kyushy some 3000 year B.C. Similarities between the two potteries strongly indicate that the Ecuadorians learned their craft from the Japanese and probably received training and implements from them over 5,000 years ago.

While there appears to be evidence to establish that Asians were in America before Columbus, more evidence was needed to determine if they came by sea or by some land bridge between continents.

In 1973 a dredge scouring the ocean floor off the coast of southern California for soil samples dredged up a doughnut-shaped stone about thirteen inches across and six inches thick, with a distinctive two-and-one-half-inch hole in its center. This item was identified as a line-weight used on ancient Asian ships. The stone was covered with a layer of manganese three millimeters thick -- a layer that might have taken natural sea processes as long as 3,000 years to deposit. The stone was identified as a fine-grained dolomite. This type stone is rarely found in California, but is abundant in northern China.

In 1975 divers made a second discovery a mile or so off the Palos Verdes Peninsula and just about 25 miles from Los Angeles. Divers found a large group of twenty stone objects. Some were doughnut shaped, some like small barrels, and some shaped like pickles. They weighed from fifty to over 1,000 pounds. Again these stones were found to be common to the southern coast of China but not of California. It is thought perhaps these stones served as anchors for ancient ships.

Some say:

These large stones are not relics from ancient ships but rather they are relics of nineteenth-century Chinese fishing vessels. It is thought that the weight of these stones was too great to be used as anchors in the waters where they were found which averaged 10-15 feet deep. And, they claim there is no fishing reef at the stone, discovery site.

Some say:

These are anchor stones made in California by Chinese fishermen who inhabited the coast in large numbers after the beginning of the California gold rush. Oddly enough one such Chinese fishing village was located near Palos Verdes.

- Could the depth of the water have changed over a period of time?
- Could the fish have been more abundant in the past than at the present time?
- Did ancient Orientals discover America long before any other cultures?

What do you think?

INANE, INNOCUOUS INVENTIONS FOR INSATIABLE INQUIRY

Inane, innocuous inventions for insatiable inquiry are zany ideas proposed and graphically portrayed to reveal complex solutions to very simple problems. Innocuous inventions are fantasy for fun with no fear of failure. Innocuous inventions challenge the mind to portray the suggested solution in an acceptable, humorous manner.

"A Simple Plant-Watering Machine"

by Ima Geneyus

A cup of water is attached to the pendulum of the clock. The cup has a hole in it near the top. When the pendulum swings out in an arc, the water level shifts, pours out the hole and drips down to the arm of the balance. The snail on the opposite end of the balance crawls out to the water droplets to get a drink. This, plus the weight of the water droplets, throws the balance out of balance and it tips downward and hits the mouse trap triggering the release of the ball. The ball is fired through the air and lands in the cup, counter balanced over a bird stand by a weighted object. The weight of the ball in the cup causes the weighted object to be pulled up. The hook attached to the weighted object engages the watering can above which, when tilted, waters the flowering plant below. Voila!

Invent an Innocuous Invention For:
- an automatic ice cream maker
- a classroom eraser cleaner
- a dog walker
- a blackboard washer
- an automatic garage door opener

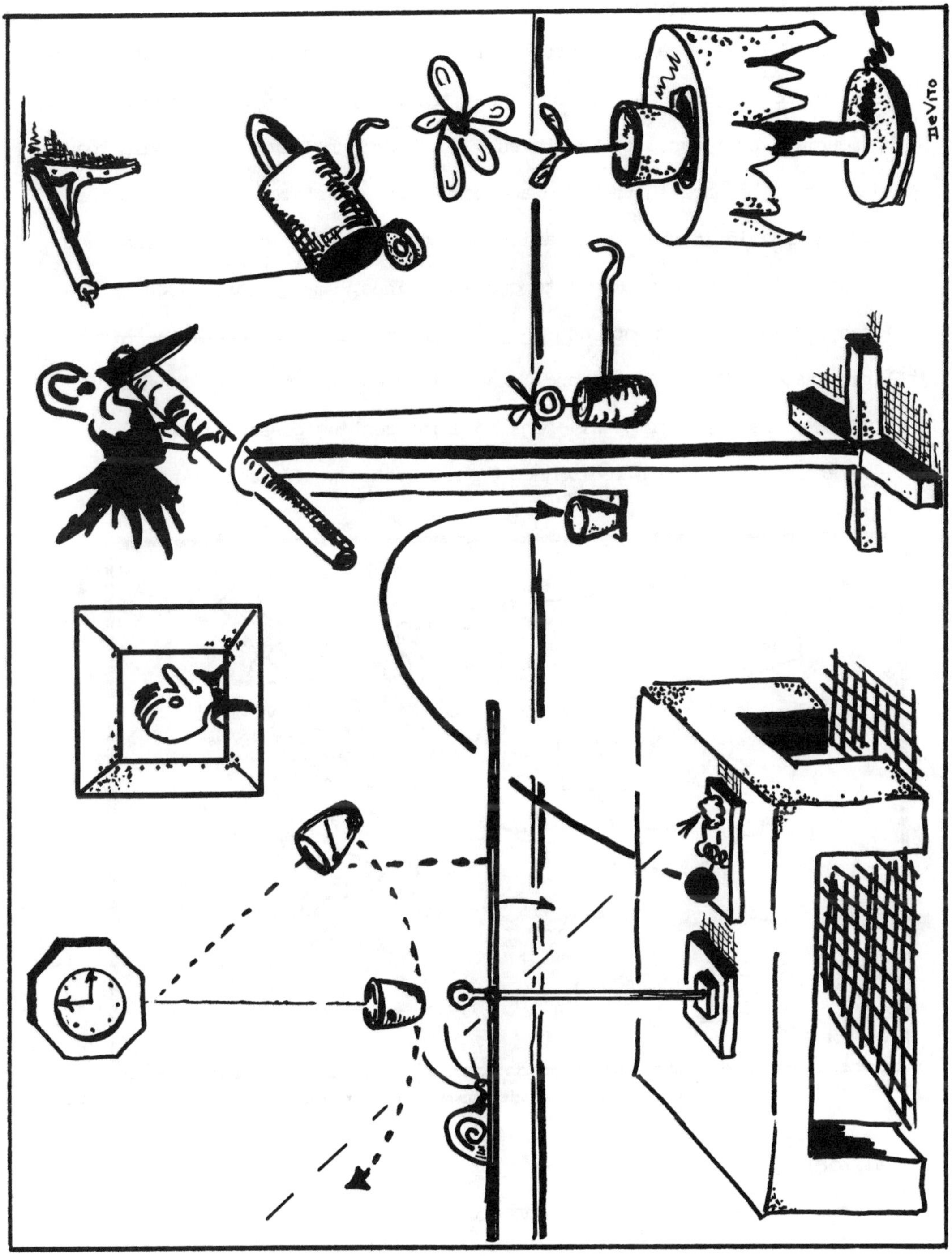
DeVito

FROM KENNEBEC TO SAN GABRIEL*

Mr. G. Etzel Pearcy, a professor of geography at California State University, Los Angeles has proposed that the 50 states in the United States be renamed and reduced to 38 states. Mr. Pearcy believes this would result in a better balance in the size and population of states. He claims that this should aid in the organization of some states and cities making them more efficient and economical to operate. Thus, small states such as New Jersey, Delaware, Vermont, and Rhode Island might cease to exist. Also, large states such as Alaska and Texas, of necessity, might be trimmed in size. Mr. Pearcy's proposal opens some interesting possibilities.

Professor Pearcy's redesigned United States would look like this:

The United States as visualized and named by G. Etzel Pearcy.

    Pearcy, G. Etzel. A 38-State United States. Plycom Press. Fullerton, California. 1974.

* From Alfred De Vito, "From Kennebec to San Gabriel" Science and Children, Vol. 14, No 2 (October, 1976), p. 39. By Permission.

336

It is estimated through this proposal that 12 states could be eliminated. Professor Pearcy estimated that this would save taxpayers $4.6 billion dollars a year in the cost of administering these 12 states. An interesting replacement for state names are drawn by Professor Pearcy from Indian names, Spanish names, and certain geologic features.

Many of the boundaries of the states, as they currently exist, were drawn up in the 18 and 19th centuries. Which state is yours? How would Professor Pearcy's proposal affect your state and you?

Did you know:

- Some state lines cut through the metropolitan areas of large cities. This leaves a portion of a major city in one state and a portion of that same city in still another state.

  > Do you have any such cities within your state?
  > What cities in the United States do you know that exist under this condition?

- Sister cities, which might better exist as one large city, are often separated by rivers, and, thus, in many instances are located in two different states.

  > What cities do you know that exist as sister cities, but that might better be organized as one city?
  > Do you have any sister cities -- one located in your state and one located in an adjoining state?

- Some states have too many people and some states have too few people, hence, they are not always equally represented in government.

  > How do you view your state - over populated, under populated or just right?

- Some states have most of the available water, other states have little water.

  > In terms of available water, is your state a have or have not state?

- Some states are occupied principally by mountainous areas and others are occupied wholly by flat land areas.

    How might you rearrange the boundary of your state to provide a more balanced geography within its borders? What does this re-arrangement do to your neighboring states?

- Some states have most of the valuable mineral wealth of the nation, others have little mineral wealth.

    Is your state a minerally rich state? If not, how would you alter your state's boundary to include mineral wealth?

- Not everyone will agree, but some states are too small for their own good and some states are too large for their own good.

    What are some of the advantages of being a large state? A small state? Name some disadvantages of being a large state? A small state?

- State boundaries are determined in a variety of ways - some by rivers, some by ocean boundaries, and some by lines of longitude and latitude.

    What conditions set the boundary of your state?

On what basis might you change the name and shape of your state? Would you agree with Professor Pearcy's proposed changes? Having thought through the problem, using the map of the United States, attempt to redesign the United States. What reasons can you give for your proposed design?

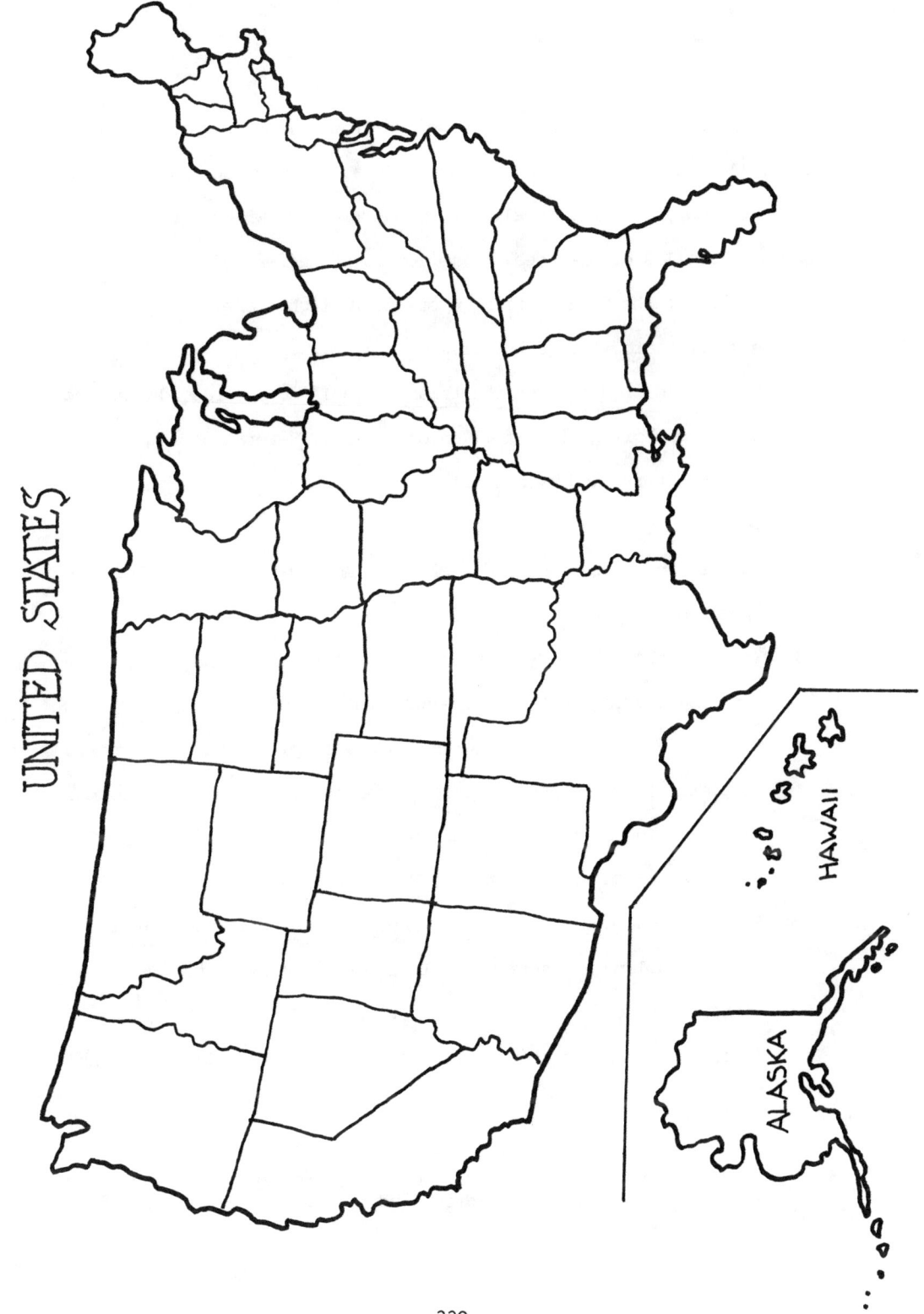

UNITED STATES

ALASKA

HAWAII

SAILOR BEWARE*

Off the coast of the continental United States lies a mysterious area of ocean. An imaginary line connecting Bermuda, Florida, and Puerto Rico forms the boundary of this so-called Bermuda Triangle, which covers about 440,000 square miles of open sea. Some lovers of hair-raising sea mysteries have called it the Deadly Triangle or the Twilight Zone.

These reports are strange but true:

1. Columbus sailed through the triangle in 1492. He recorded seeing a strange bolt of fire fall into the sea. Several days later his men were terrified by a baffling disturbance of the ship's compass.

2. Between the years 1781 and 1812, four American naval vessels disappeared without explanation.

3. In 1918, the U.S.S. Cyclops sailed northward into the triangle -- and disappeared. Not one clue was ever found.

4. Since 1946, some twenty other ships and airplanes have disappeared completely. One strange reported case involved five Navy torpedo bombers. Inside the deadly triangle, the patrol leader radioed, "Strange things are happening. We can't be sure of any direction. Even the sea doesn't look as it should." Then all was silent. A rescue seaplane was sent out. In twenty minutes it too vanished without a trace.

5. In 1963, a merchant ship, a fishing boat, and two Air Force tanker jets disappeared.

---

* From Alfred De Vito and Gerald H. Krockover, <u>Creative Sciencing: Ideas and Activities for Teachers and Children</u>, 2nd ed. p. 89. Copyright © 1980 by Alfred De Vito and Gerald H. Krockover. Reprinted by permission of Little, Brown and Company.

Though many ships and planes travel this triangle each day, an unusual number of craft have vanished without a trace. Neither wreckage nor victims were ever found.

What explanations can you suggest to account for these strange events?

How might you set out to support your explanations?

Would you travel through or over the Bermuda Triangle?

Is this rate of losses natural for every 440,000 square miles of open sea?

Is anything unusual about the position of this triangle on the earth's surface that might account for these strange events?

THE DELICATE DILEMMA[*]

If the current quality of life on earth is to be maintained, it has been stated that zero population growth at the 7 billion level is required. With controls, by the turn of the century the world should reach a stable population of 7 billion people. Without controls, the population of the world by the year 2000 might reach close to 35 billion people.

With the recent major accomplishments in the space program, some people have suggested that a solution to the world's over-population is the colonization of space. It has been suggested that we simply prepare to move out to another planet in our own galaxy or another galaxy.

By contrast:

The United States space vehicle Eagle III is to land on a nearby planet next summer. One of its assignments is to test soil samples on the spot for signs of life. Life would be indicated by the presence of such organic compounds as carbon dioxide and methane.

If evidence of life is found, world biologists would be most anxious to compare life on this planet to earth life in the laboratory.

The problem:

Scientists are arguing whether to keep or kill living things a spacecraft eventually may bring back from other places in space.

Some thoughts:

Living things brought in from space may infect the earth and endanger all of us.

Some scientists believe living organisms from other places in space

---

[*] From Alfred De Vito and Gerald H. Krockover, Creative Sciencing: Ideas and Activities for Teachers and Children, 2nd edition, p. 92. Copyright © 1980 by Alfred De Vito and Gerald H. Krockover. Reprinted by Permission of Little, Brown and Company.

would be so affected by the earth's temperature, humidity, and oxygen content that they would be difficult to keep alive, even in a laboratory. Thus, they doubt that any living thing brought to earth would present a threat of infection.

Some scientists believe that what we might learn from living things brought from outer space would be so valuable, we should take the risk.

Other scientists believe that the threat from unknown danger is so great that no risk should be taken.

Are there solutions to these problems? Should we colonize space?
If earth people invade space, might we infect the planets we land on?
Should we permit living things from outer space to invade earth?
What do you think?

The Finale

MEMORABLE, MAGNANIMOUS NUGGETS FOR MAGNIFICENT, MELLIFLUENT SCIENCE TEACHING

<u>Methodology</u>:

- Start simple - complexity will catch up to you.
- Everytime you improve your teaching, you
    increase your work load...however,
    this is usually matched by accompanying
    improved body strength.
- In teaching one gets the respect and admiration
    one deserves, and sometimes a
    little extra.
- Be instructionally oriented at all times.
    Know where you have been (previous lessons).
    Know where you are going (today's lesson).
    And, know when you have arrived (evaluation).
    And, within this structure remain like a
    chip on the water ready to grasp at those
    spontaneous "teachable" moments.
- Creativity enhances instruction and learning;
    like Archimedes' spiral creativity moves
    ideas to the surface.
- Personal attitudes and desires will win out
    over mandated learning every time.

- Praise, fun, games, interesting involvements, sprite, bright, light, scintillating mind captures are highly valued components of science instruction. They become more valuable when coupled with an acquired desire by children to enjoy work and practice it in a disciplined manner. Children should like what they do, but they must also learn to like what they must do.
- Children need to know what the expectations are. They need to know what game you are playing and what the rules are in order to play.
- Nothing works like authors claim, at least not for the first, three attempts. They provide you with threads and you must weave the fabric.

Science Instruction:
- A great science program poorly presented is a poor program. A poor or an average program in the hands of an outstanding teacher can be an outstanding program.
- There is no such thing as elementary science - there is science for elementary school children.
- Science is rooted in observation. Instruction in observation must be continuous; refinement in this process is endless.
- Curiosity is not enough, better a burning desire to know.

- Classification is a natural outcome of the act of observing.
- Where applicable quantify. If something moves, record how fast it moved. How far it moved. And, in what direction it moved. Who knows when the data will be needed.
- Safety concerns in science instruction dictate that when you are in doubt about the safety of something, "leave it out." Good prudent judgement will always stand you in good stead.

<u>Problem Solving</u>:
- There isn't one best way to solve problems. There are, however, preferred ways.
- For every problem an answer or explanation exist, but one may never find it; we may, for the moment, need to settle for simply getting closer to the answer.
- Provocative questioning doesn't always solve problems, but it can move one closer to the solution.

Index

Assumption, 10

Biology, opportunities
    for investigation, 92,
    97, 100, 103
Botany, opportunities for
    investigation, 102, 104
Brainstorming, 105

Challenges for Children, 131
Chemistry, opportunities for
    investigation, 102
Classifying, 31
    Activities, 63
Communication, 31
Controlling variables, 34
Controlling variables
    activities, 81
Creativity, barriers to
    teaching, 125
Creativity, 117
Creativity, components, 119
    And questioning, 139
    Improvement of, 123
Creative Involvements, 145
    Model, 119
    Persons, attributes
        of, 120
Crystal systems, 215-221

Dilemma
    Science activities
        (Index), 264
    Analysis attributes, 268
    Analysis preparation, 270
Discrepant events, 228
Divergent-convergent model, 121
Electroculture, opportunities
    for investigation, 94
Extrapolation, 75
Experimentation, 4, 14, 21, 85
Experimenting, 35
    Activities, 85

Finale, 344

Generalization, 9
Geology, opportunities for
    investigation, 98

Graphing, 32
    Activities, 72
    Bar graph, 74
    Direct proportion, 73
    Histogram, 74
    Inverse proportion, 73
    Slope, 76

Hathaway Attawhey, 142
Health, opportunities for
    investigation, 104
Horticulture, opportunities for
    investigation, 92, 93, 96
Hypothesis, 9
Hypothesizing, 34
    Activities, 85

Ideational-Generation approach,
    105
Inference, 10
Inferring, 30
    Activities, 42, 54
Interpolation, 75

Law, 9
Lesson plan, gifted, 132

Match stick problems, 185, 186
Model, 10
    Analogue, 148
    Abstract, 148
    Building, 147
    Concrete, 148
    For dissolving, 161
    For spatial development, 179
    Umbrellic Science, 157
Morphological approach, 18
Mystery containers, 150

Observing, 28
Observation activities, 42, 44
Observation, 9
Oragami, 203, 204, 205

Physics, opportunities for
    investigation, 93, 95, 96,
    101
Prediction, 30
Predicting activities, 43, 58

Process approach, 27
    Skills, 27
Provocative questions, 144
Puzzles and problems, 257

Questions, how and why, 146

Science
    Acquired individual
      attributes, 4
    Applied, 1
    Content selection, 7
    Definition, 3
    Desirable outcomes, 5
    For the gifted, 128
    Formula, 11
    The language, 8
    Methodology, 5
    Pure, 1
Space/Time relationships, 33
Stimulating thinking, 134

Tangrams, 182, 183, 184
Taxonomy of educational
    objectives, 139
Tessellations, 194, 195, 196
Theory, 10

Using numbers, 32

Variables
    Controlled, 80
    Manipulated, 80
    Responding, 80

What would happen if..., 143